LONDON MATHEMATICAL SOCIETY LECTURE NOTE SERIES

This series publishes the records of lectures and seminars on advanced topics in mathematics held at universities throughout the world. For the most part, these are at post-graduate level either presenting new material or describing older matter in a new way. Exceptionally, topics at the undergraduate level may be published if the treatment is sufficiently original.

Prospective authors should contact the editor in the first instance.

Already published in this series

1. General cohomology theory and K-theory, PETER HILTON.
4. Algebraic topology: A student's guide, J.F. ADAMS.
5. Commutative algebra, J.T. KNIGHT.
7. Introduction to combinatory logic, J.R. HINDLEY, B. LERCHER and J.P. SELDIN.
8. Integration and harmonic analysis on compact groups, R.E. EDWARDS.
9. Elliptic functions and elliptic curves, PATRICK DU VAL.
10. Numerical ranges II, F.F. BONSALL and J. DUNCAN.
11. New developments in topology, G. SEGAL (ed.).
12. Symposium on complex analysis Canterbury, 1973, J. CLUNIE and W.K. HAYMAN (eds.).
13. Combinatorics, Proceedings of the British combinatorial conference 1973, T.P. McDONOUGH and V.C. MAVRON (eds.).
14. Analytic theory of abelian varieties, H.P.F. SWINNERTON-DYER.
15. An introduction to topological groups, P.J. HIGGINS.
16. Topics in finite groups, TERENCE M. GAGEN.
17. Differentiable germs and catastrophes, THEODOR BRÖCKER and L. LANDER.
18. A geometric approach to homology theory, S. BUONCRISTIANO, C.P. ROURKE and B.J. SANDERSON.
19. Graph theory, coding theory and block designs, P.J. CAMERON and J.H. VAN LINT.
20. Sheaf theory, B.R. TENNISON.
21. Automatic continuity of linear operators, ALLAN M. SINCLAIR.
22. Presentations of groups, D.L. JOHNSON.
23. Parallelisms of complete designs, PETER J. CAMERON.
24. The Topology of Stiefel manifolds, I.M. JAMES.
25. Lie groups and compact groups, J.F. PRICE.
26. Transformation groups: Proceedings of the conference in the University of Newcastle upon Tyne, August 1976, CZES KOSNIOWSKI.

continued overleaf

27. Skew field constructions, P.M. COHN.
28. Brownian motion, Hardy spaces and bounded mean oscillation, K.E. PETERSEN.
29. Pontryagin duality and the structure of locally compact abelian groups, SIDNEY A. MORRIS.
30. Interaction models, N.L. BIGGS.
31. Continuous crossed products and type III von Neumann algebras, A. VAN DAELE.
32. Uniform algebras and Jensen measures, T.W. GAMELIN.
33. Permutation groups and combinatorial structures. N.L. BIGGS and A.T. WHITE.
34. Representation theory of Lie groups, M.F. ATIYAH.
35. Trace ideals and their applications, BARRY SIMON.
36. Homological group theory, edited by C.T.C. WALL.
37. Partially ordered rings and semi-algebraic geometry, GREGORY W. BRUMFIEL.

London Mathematical Society Lecture Note Series. 38

Surveys in Combinatorics

Edited by B. Bollobás

Proceedings of the
7th British Combinatorial Conference

Cambridge University Press
Cambridge
London New York Melbourne

CAMBRIDGE UNIVERSITY PRESS
Cambridge, New York, Melbourne, Madrid, Cape Town, Singapore, São Paulo

Cambridge University Press
The Edinburgh Building, Cambridge CB2 8RU, UK

Published in the United States of America by Cambridge University Press, New York

www.cambridge.org
Information on this title: www.cambridge.org/9780521228466

First published 1979
Re-issued in this digitally printed version 2008

A catalogue record for this publication is available from the British Library

ISBN 978-0-521-22846-6 paperback

Contents

Preface

Since its inception at Oxford in 1969 the British Combinatorial Conference has become a regular feature of the international mathematical calendar. This year the seventh conference will be held in Cambridge from 13th to 17th August, under the auspices of the Department of Pure Mathematics and Mathematical Statistics. The participants and the contributors represent a large variety of nationalities and interests.

The principal speakers were drawn from the mathematicians of Britain, Europe and America. They were asked to review the diverse areas of combinatorics in which they are expert. In this way it was hoped to provide a valuable work of reference describing the state of the art of combinatorics. All of the speakers kindly submitted their articles in advance enabling them to be published in this volume and made available in time for the conference.

I am grateful to the contributors for their cooperation which has made my task as an editor an easy one. I am also grateful to the Cambridge University Press, especially Mr David Tranah, for their efficiency and skill. On behalf of the British Combinatorial Committee I would like to thank the British Council, the London Mathematical Society and the Mathematics Faculty of Cambridge University for their financial support.

June 1979 Béla Bollobás

1 · Resonance and reconstruction

N.L. BIGGS

SUMMARY

This article is another attempt to reconcile part of graph theory and part of theoretical physics. Specifically, we shall discuss some aspects of the reconstruction problem in terms of simple models of physical phenomena. A previous essay in the same vein (Biggs 1977a, henceforth referred to as IM) may be consulted for background information and proofs of some basic theorems.

There are four sections: (1) Interaction Models, (2) the Algebra of Graph Types, (3) Reconstruction, (4) Partition Functions for Infinite Graphs.

1. INTERACTION MODELS

1.1 Definitions

Let G be a finite (simple) graph, with vertex set V and edge-set E. An 'interaction model' on G arises when the vertices of G may have certain attributes, and they interact along the edges according to the values of those attributes. To be more precise, let A denote a finite set of objects (attributes), and define a *state* on G to be a function $\omega: V \to A$. In graph theory, we often think of the attributes as colours, so that a state is a colouring. In theoretical physics, the vertices represent particles of some kind, and the attributes represent some physical property, such as a magnetic moment.

The interaction between a pair of adjacent vertices is measured by a real-valued function i, called the *interaction function*, defined on the set $A^{(2)}$ of unordered pairs of attributes. That is, $i\{a_1,a_2\}$ measures the interaction between a pair of vertices when they are joined by an edge and they have attributes a_1 and a_2. An *interaction model* M is a pair (A,i), where A is a finite set and i is an interaction function. Thus different interaction models may represent different physical contexts, while graphs represent possible geometrical configurations of particles to which the physical laws apply.

Suppose that an interaction model M and a graph G are given. The total 'weight' $I(\omega)$ of a state ω on G is defined to be

$$I(\omega) = \prod_{\{u,v\}\in E} i\{\omega(u),\omega(v)\} .$$

(The reason for taking a product, rather than a sum, will be explained shortly.) The *partition function* for the model M on G is

$$Z(M,G) = \sum_{\omega:V\to A} I(\omega) . \qquad (1.1.1)$$

1.2 Examples

In models of physical phenomena, the weight $I(\omega)$ is usually represented by an expression of the form $\exp[H(\omega)/kT]$, where k is an absolute constant, T is the temperature, and $H(\omega)$ is a measure of the energy of the state ω. Thus every state has a positive weight, and $I(\omega)/Z$ represents the probability of finding the system in a state ω. The expression for $H(\omega)$ will be a sum of terms $X(e)$, one for each 'local interaction' (edge) of the system; thus

$$I(\omega) = \exp \frac{1}{kT} \sum_{e\in E} X(e) = \prod_{e\in E} \exp X(e)/kT .$$

So we have recovered the general formulation for $I(\omega)$, and justified the occurrence of the product. In general, we think of the partition function Z as the representative of the global properties of the system.

A specific example is the famous Ising model of magnetism. In this case, the vertices of the graph represent particles in a ferromagnetic substance, and the attributes are the two possible orientations of a magnetic moment, conventionally described as 'up' and 'down'. If two adjacent particles have the same orientation, they contribute an amount of energy $+L$ to $H(\omega)$, otherwise they contribute $-L$. In order to express this model in our general framework, we may replace 'up' and 'down' by 0 and 1, and define an interaction function for $A = \{0,1\}$ by

$$i\{0,0\} = i\{1,1\} = \varepsilon \quad , \qquad i\{0,1\} = \varepsilon^{-1} \quad ,$$

where $\varepsilon = \exp(L/kT)$. We shall denote this interaction model (A,i) by I_T and refer to it as the *Ising model at temperature* T.

An example more familiar to graph theorists is the *colouring model* C_u. Here A is a finite set of u $(= |A|)$ colours, and the interaction function is defined as follows:

$$i\{a_1,a_2\} = \begin{cases} 0 & \text{if } a_1 = a_2 \ , \\ 1 & \text{otherwise} \ . \end{cases}$$

In this model, the weight of a 'colouring' $\omega: V \to A$ of the vertices of a graph G is zero if any pair of adjacent vertices have the same colour, and 1 otherwise. Hence $Z(C_u,G)$ is simply the number of proper colourings of G when u colours are available.

1.3 Resonant Models

Both the Ising model I_T and the colouring model C_u have the property that the interaction function takes only two values: $i\{a_1,a_2\}$ has one value i_0 if $a_1 = a_2$, and another value i_1 in all other cases. We may think of such a model as representing a situation where two particles interact in a special way if they have the same attributes, and in some other (constant) way if not. This is a kind of resonance, and we shall refer to such an interaction model as a *resonant* model.

The definition (1.1.1) of the partition function shows that, if R is a resonant model, $Z(R,G)$ is essentially dependent only on the ratio i_0/i_1. For this reason, we shall confine our attention to a normalized resonant model R_β whose interaction function is given by

$$i\{a_1,a_2\} = \begin{cases} \beta & \text{if } a_1 = a_2 , \\ 1 & \text{otherwise} . \end{cases} \qquad (1.3.1)$$

The following result is perhaps the most important property of resonant models.

<u>Theorem A</u> Let $R_\beta = (A,i)$ be a resonant model with $|A| = u$ and interaction function i as in (1.3.1), and let $G = (V,E)$ be a graph. Then

$$Z(R_\beta,G) = u^{|V|} \sum_F (\beta-1)^{|F|} u^{-r(F)} \qquad (1.3.2)$$

The sum is taken over all subsets F of E, and $r(F)$ denotes the rank of the edge-subgraph $\langle F\rangle$.

<u>Proof</u> [IM, p.23].

Since $r(F) < |V|$ for all subsets F of E, Theorem A tells us that the partition function of a resonant model is a polynomial function of the number of attributes, u. In the case $\beta = 0$ we obtain the well-known chromatic polynomial,

$Z(C_u, G)$. The significance of the polynomial property in the general theory will appear in Section 3.

It may be noted that the polynomial property, and the resonance property, are both related to the existence of a 'deletion and contraction' algorithm: see Vout (1978).

2. THE ALGEBRA OF GRAPH TYPES

2.1 Star Types and Graph Types

A *star type* is an isomorphism class of finite, simple, non-separable graphs. We shall use the symbol St to denote the set of star types. It is often convenient to use a pictographic representation for the smaller star types: St = {|,△,□,⧄,...} .

A *graph type* is a function t defined on St and taking non-negative integer values, only a finite number of which are non-zero. We shall use the symbol Gr to denote the set of graph types. A finite graph G has *type* t if, for each $\sigma \in$ St , G has $t(\sigma)$ blocks of star type σ . Thus two graphs of the same type are not necessarily isomorphic; however, we shall see that the equivalence relation 'of the same type' is the appropriate one for the study of interaction models.

We shall denote the vector spaces of complex-valued functions defined on St , and on Gr , by $\underset{\sim}{X}$ and $\underset{\sim}{Y}$ respectively. Since St may be regarded as a subset of Gr in the obvious way, we have a projection $J: \underset{\sim}{Y} \to \underset{\sim}{X}$ defined by $(Jy)(\sigma) = y(\sigma)$ $(\sigma \in \text{St})$.

2.2 Type-Invariants

A function f defined on the set of finite graphs is a *type-invariant* if $f(G_1) = f(G_2)$ whenever G_1 and G_2 have the same type. In order that f may be a type-invariant, it is clearly sufficient that it should be an isomorphism invariant (so that it is invariant for star types), and that it should be multiplicative over blocks:

$$f(G) = \Pi f(B) \quad ,$$

where the product is taken over the set of blocks B of G.

The partition function of an interaction model is not quite a type-invariant, since the multiplicative property does not hold. (The partition function for two disjoint blocks is not the same as that for two blocks with one common vertex.) However, it is easy to see how to remedy this difficulty. We define the *reduced* partition function of the model $M = (A,i)$, with $|A| = u$, to be

$$\bar{Z}(M,G) = Z(M,G)/u^{|V|} \quad .$$

Theorem B The reduced partition function $\bar{Z}(M,G)$ is a type-invariant.

Proof [IM, p.63].

Associated with any type-invariant function f there is a vector ϕ in $\underset{\sim}{Y}$ defined by $\phi(t) = f(T)$, where T is any graph of type t. Thus we may think of the reduced partition function, with respect to a given model, as an element of $\underset{\sim}{Y}$.

2.3 Counting Subgraphs

If s and t are graph types, we define c_{st} to be the number of edge-subgraphs of a graph S of type s which have type t. There are two ways of fitting these numbers into our algebraic framework. First, we may think of the array (c_{st}) as a matrix, so that we have a linear transformation $C: \underset{\sim}{Y} \to \underset{\sim}{Y}$, defined as follows:

$$(Cy)(s) = \sum_t c_{st}\, y(t) \quad . \qquad (2.3.1)$$

It is clear that, for each given s, the sum on the right-hand side involves only a finite number of non-zero terms c_{st}. Furthermore:

<u>Theorem C</u> The linear transformation C is invertible.

<u>Proof</u> We have only to notice that if the graph types are ordered in a suitable way (for instance, compatibly with increasing number of edges) then the matrix (c_{st}) is lower triangular, and its diagonal terms are non-zero. Hence the terms of an inverse matrix may be computed recursively in the usual way.

Another useful way of handling the numbers c_{st} is to define, for each graph type s, a vector c_s in $\underset{\sim}{Y}$ as follows:

$$c_s(t) = c_{st} \quad . \qquad (2.3.2)$$

The vector c_s, giving the census of subgraphs of s, may be considered as the representative of a 'real' graph type s. The result of Theorem C implies that the vectors $\{c_s\}$ $(s \in Gr)$ form a basis for $\underset{\sim}{Y}$, so that each y in $\underset{\sim}{Y}$ may be expressed (uniquely) as a linear combination of the basis $\{c_s\}$. Thus y is a 'generalized' graph type.

The point of view developed in the previous paragraph was introduced by Whitney in his pioneering work on graph colouring. He noticed that c_s is determined by its projection Jc_s in $\underset{\sim}{X}$; more precisely:

<u>Theorem D</u> There is a (non-linear) operator $W: \underset{\sim}{X} \to \underset{\sim}{Y}$, independent of the graph type s, such that

$$W(Jc_s) = c_s \quad (\text{for all } s \in Gr) \quad .$$

Proof [IM, p.67]. (The first proof (Whitney, 1932) was rather complicated; he failed to invert a matrix. Fortunately, I had the help of Colin Vout, who did.)

2.4 Expansions in Algebraic Form

The polynomial expansion (1.3.2) for the partition function of a resonant model may be written in algebraic form. For convenience we introduce a new variable $z = 1/u$, where $u = |A|$ is the number of attributes, and use the reduced partition function. The formula (1.3.2) becomes:

$$\bar{Z}(R_\beta, G) = \sum_{F \subseteq E} (\beta - 1)^{|F|} z^{r(F)} \quad . \tag{2.4.1}$$

The reduced partition function and the individual summands on the right-hand side are type-invariants. Thus, if we write $e(s)$ and $r(s)$ for the number of edges and the rank of a graph of type s, we may define vectors ρ_z and m_z in $\underset{\sim}{Y}$ as follows:

$$\rho_z(s) = \bar{Z}(R_\beta, S) \quad , \quad m_z(s) = (\beta - 1)^{e(s)} z^{r(s)} \quad ,$$

where S is any graph of type s. If we now collect the terms in (2.4.1) according to the type of the subgraph $\langle F \rangle$, we obtain

$$\rho_z(s) = \sum_t c_{st} m_z(t) \quad .$$

Equivalently, $\rho_z = Cm_z$. In fact, this expression is quite general, and does not depend on the resonance property of the model. If ξ is the vector representing the reduced partition function of an interaction model M, then the invertibility of the transformation C ensures that there is a vector m in Y such that $\xi = Cm$. That is,

$$\xi(s) = \sum_t c_{st} \, m(t) \quad . \tag{2.4.2}$$

It is helpful to interpret (2.4.2) in the following way: each subgraph of type t contributes an amount $m(t)$ to the value of $\xi(s)$. In the case of a resonant model, we have the useful feature that the contributions are simple monomial expressions in β and u .

3. RECONSTRUCTION

3.1 The Reconstruction Problem

In this section we shall explain the relationship between the foregoing ideas and the 'reconstruction problem' in graph theory. The basic theory was first published by Tutte (1967); there is also an account of it in Biggs (1974). However, it was not until Tutte's more recent work became available that its relevance was generally recognised (Tutte, 1979).

Let G be a finite simple graph with vertex-set $V = \{v_1, \ldots, v_n\}$, and let G_i denote the vertex-subgraph $\langle V - v_i \rangle$ of G . (G_i is obtained from G by deleting v_i and the edges incident with it.) In the *vertex-reconstruction problem* we are given the set of graphs $\{G_1, \ldots, G_n\}$, unlabelled and unordered, and we ask how much information about G may be deduced: such information is said to be *reconstructible*. It is possible that G itself may be reconstructible, but in general this seems to be a difficult question. We shall show that the partition function $Z(\mathcal{R}, G)$ is reconstructible, for any resonant model $\mathcal{R}$.

We begin by taking a census of the non-separable vertex-subgraphs of G . For a given graph S of type s , and a given star type τ , define $k_{s\tau}$ to be the number of vertex-subgraphs of S which have type τ . Now each non-separable vertex-subgraph $\langle W \rangle$ of G is a vertex-subgraph of those G_i for which v_i is not in W. Thus if g denotes the type of G, and H denotes the family of types of $G_1, \ldots, G_n$, we have

$$\{|V| - v(\tau)\} k_{g\tau} = \sum_{h \in H} k_{h\tau} , \tag{3.1.1}$$

where $v(\tau)$ is the number of vertices of the star type τ. From this, we deduce the following useful result.

<u>Theorem E</u> Suppose that σ and τ are two star types which occur as vertex-subgraphs of a graph G. The the number $k_{\sigma\tau}$ is reconstructible.

<u>Proof</u> Clearly, there is nothing to prove unless σ is the type of G. In that case, $k_{\sigma\sigma}$ is unity, and the other values of $k_{\sigma\tau}$ are determined by (3.1.1).

For example, let G be a graph of type ⧄. Then $H = \{||, ||, \Delta, \Delta\}$ and we have

$$(4-2) k_{⧄, |} = 2 + 2 + 3 + 3 ,$$

and so forth. The full set of relevant values of $k_{\sigma\tau}$ may be tabulated as follows:

	|	Δ	⧄
|	1	0	0
Δ	3	1	0
⧄	5	2	1

(3.1.2)

3.2 The Multiplicative Method

Let us suppose that the star types have been listed in an order compatible with increasing number of vertices. The matrix $(k_{\sigma\tau})$ becomes lower triangular, and each diagonal term is 1. (For a given graph, the relevant values of $k_{\sigma\tau}$ form a finite submatrix, as for example in (3.1.2).) It follows that $(k_{\sigma\tau})$ has an inverse matrix $(\ell_{\sigma\tau})$, and the numbers $\ell_{\sigma\tau}$ are integers (positive, negative, or zero).

Let $\mathcal{R} = (A,i)$ be a resonant model, and let $z = 1/u$, where $u = |A|$. There is an associated ρ_z in $\underset{\sim}{X}$ defined by

$$\rho_z(\sigma) = \bar{Z}(\mathcal{R},G) \quad ,$$

where G is any graph of type σ. Furthermore, Theorem A tells us that $\rho_z(\sigma)$ is a polynomial function of z, and it has the form

$$\rho_z(\sigma) = 1 + B_1 z + B_2 z^2 + \ldots + B_r z^r \quad ,$$

where r is the rank of σ, and the coefficients $B_1,\ldots,B_r$ are functions of the interaction values. Henceforth we shall assume that the definition of $\rho_z(\sigma)$ has been extended to all complex values of z by means of the same polynomial expression. We define

$$\pi_z(\sigma) = \Pi \{\rho_z(\tau)\}^{\ell_{\sigma\tau}} \quad . \tag{3.2.1}$$

The product is taken over all star types τ for which $\ell_{\sigma\tau} \neq 0$, and since these numbers are integers, $\pi_z(\sigma)$ is a rational function of z.

<u>Theorem F</u> If ρ_z and π_z are defined as above, we have

$$\rho_z(\sigma) = \Pi \{\pi_z(\tau)\}^{k_{\sigma\tau}} . \qquad (3.2.2)$$

Proof Substitute the values of π_z (3.2.1) in the right-hand side, and recall that $(k_{\sigma\tau})$ and $(\ell_{\sigma\tau})$ are inverse matrices.

Theorem F says that the value of $\rho_z(\sigma)$ may be obtained by multiplying factors $\pi_z(\tau)$, one for each vertex-subgraph of σ which has type τ . For example,

$$\rho_z(\boxtimes) = \pi_z(|)^5 \pi_z(\Delta)^2 \pi_z(\boxtimes)^1 .$$

At first sight, this seems rather unpromising. The formula for $\rho_z(\sigma)$ contains a factor $\pi_z(\sigma)$, and the definition of $\pi_z(\sigma)$ contains a factor $\rho_z(\sigma)$. In order to overcome this difficulty, we need some more facts about π_z , which we shall state here and prove in the next subsection.

We have already remarked that $\pi_z(\sigma)$ is a rational function of z , and so its singularities in the complex plane are isolated poles. The origin is not a pole, since $\rho_0(\tau) = 1$ for all $\tau \in$ St . Hence $\pi_z(\sigma)$ has a Taylor series expansion at $z = 0$, with a non-zero radius of convergence.

Theorem G Suppose the Taylor expansion of $\pi_z(\sigma)$ at $z = 0$ is

$$\pi_z(\sigma) = \sum_{i=0}^{\infty} p_i z^i .$$

Then we have

$$p_0 = 1 , \quad p_i = 0 \quad (1 \le i < v(\sigma) - 1) .$$

Proof See subsection 3.3.

We now have all the machinery needed to show that $\bar{Z}(R,G)$ is reconstructible. We shall explain the method by means of a simple example, taking G to be a graph of type $\boxslash$ and R to be the colouring model C_u. First, we recall that the portion of the matrix $(k_{\sigma\tau})$ relevant to G is reconstructible, and that for a graph of type $\boxslash$ it is given by (3.1.2). We shall calculate $\rho_z(\boxslash) = \bar{Z}(R,G)$ recursively, using (3.1.2) and the following properties of ρ_z and π_z :

(P1) $\rho_z(\sigma)$ is a polynomial of degree $v(\sigma) - 1$;

(P2) $\rho_z(\sigma)$ satisfies the multiplicative formula (Theorem F);

(P3) $\pi_z(\sigma)$ has 'vanishing coefficients' (Theorem G);

(P4) $\rho_1(\sigma) = 0$ (from the definition of $\bar{Z}$ with $z = u = 1$).

The method involves the calculation of ρ_z and π_z for each subgraph in turn. Thus:

<u>Step 1</u> $\rho_z(|) = 1 + az$ (P1) and $\rho_1(|) = 0$ (P4) .

Hence $\rho_z(|) = 1 - z$ and $\pi_z(|) = 1 - z$ (P2) .

<u>Step 2</u> $\rho_z(\Delta) = \pi_z(|)^3 \pi_z(\Delta) = (1-z)^3(1 + bz^2 + \ldots)$ (P2,P3)

$= 1 - 3z + cz^2$ (P1) .

$\rho_1(\Delta) = 0$ (P4) , hence $c = 2$,

$\rho_z(\Delta) = 1 - 3z + 2z^2$ and $\pi_z(\Delta) = (1-2z)/(1-z)^2$ (P2) .

<u>Step 3</u> $\rho_z(\boxslash) = \pi_z(|)^5 \pi_z(\Delta)^2 \pi_z(\boxslash)$ (P2)

$= (1-z)(1-2z)^2(1 + dz^3 + \ldots)$ (P3)

$= 1 - 5z + 8z^2 + ez^3$ (P1) .

$\rho_1(\boxslash) = 0$ (P4) , hence $e = -4$ and

$\rho_z(\boxslash) = 1 - 5z + 8z^2 - 4z^3$.

This method works for any non-separable graph. If G has more than one block then we simply use the fact that $\bar{Z}$ is multiplicative over blocks. For a general resonant model R_β (1.3.1), the same method applies, except that (P4) becomes $\rho_1(\sigma) = \beta^{e(\sigma)}$, where $e(\sigma)$ is the number of edges of σ .

Thus we see that $\bar{Z}(R_\beta, G)$ is reconstructible.

3.3 Star-Clusters and Vanishing Coefficients

In this subsection we shall sketch a proof of Theorem G, using the algebra of graph types. This proof should be compared with Tutte's (1979), which uses formal power series.

Let S be a finite simple graph of type s, with edge-set E. A set $S = \{E_1, \ldots, E_r\}$ of distinct subsets of E is a *star-cluster* in S if each of the edge-subgraphs $\langle E_i \rangle$ is non-separable. The function t_S defined by the rule that $t_S(\sigma)$ is the number of $\langle E_i \rangle$ which have type σ, is called the *type* of S. We say that S *covers* S if the union of the sets E_i is E.

Let h_{st} denote the number of star-clusters σ in S for which $t_S = t$, and let v_{st} denote the number of such star-clusters which cover S. We have the following relations between h_{st}, v_{st} and the numbers c_{st} previously defined.

Theorem H

$$h_{st} = \prod_{\tau \in St} \binom{c_{s\tau}}{t(\tau)} , \tag{3.3.1}$$

$$h_{st} = \sum_{q \in Gr} c_{sq} v_{qt} . \tag{3.3.2}$$

Proof By counting.

Now let R be a fixed resonant model and ρ_z the associated function defined in Section 3.2. Let (d_{st}) be the (integral) matrix inverse to (c_{st}) and define

$$\lambda_z(\sigma) = \prod_\tau \{\rho_z(\tau)\}^{d_{\sigma\tau}} \quad (\sigma, \tau \in St) . \tag{3.3.3}$$

In the following arguments we shall assume that the complex number z has been chosen so that certain expressions involving λ_z are meaningful. The justification for this is that $\lambda_z(\sigma)$ is defined in a neighbourhood of $z = 0$, and the expressions concerned involve only a finite number of terms $\lambda_z(\sigma)$. With this in mind, we define

$$\kappa_z(t) = \Pi \{\lambda_z(\tau) - 1\}^{t(\tau)} \qquad (t \in Gr) , \qquad (3.3.4)$$

the product being taken over the set of star types τ for which $t(\tau) \neq 0$.

If we are given a graph type s, the numbers h_{st} and v_{st} are non-zero for only finitely many types t. Hence there are linear mappings $H,V: Y \to Y$, defined by $(Hy)(s) = \Sigma h_{st} y(t)$, $(Vy)(s) = \Sigma v_{st} y(t)$. Furthermore, a simple calculation based on the distributive law [IM, p.95] and equation (3.3.1) shows that $H\kappa_z = \rho_z$.

We now recall that $\rho_z = Cm_z$ (2.4.1) and $H = CV$ (3.3.2). Thus $CV\kappa_z = Cm_z$, and since C is invertible,

$$V\kappa_z = m_z . \qquad (3.3.5)$$

The matrix (v_{st}) is upper triangular, with non-zero diagonal terms (and essentially finite rows), if we adopt our usual convention about ordering the graph types. Thus an inverse matrix can be found, also upper triangular, although its rows are not essentially finite. In particular, for any given type σ, we may invert (3.3.5) to obtain an expression for $\kappa_z(\sigma)$ as an infinite series of terms $m_z(t)$. Furthermore we know that $m_z(t) = (\beta-1)^{e(t)} z^{r(t)}$, and the only terms with non-zero coefficients have $r(t) \geq r(\sigma)$. Hence

$$\lambda_z(\sigma) = 1 + \kappa_z(\sigma) = 1 + p_r z^r + p_{r+1} z^{r+1} + \dots ,$$

where $r = r(\sigma) = v(\sigma) - 1$. To obtain Theorem G we must translate this into a result about vertex-subgraphs. Now each edge-subgraph of a graph G is a spanning subgraph of a unique vertex-subgraph of G. Hence, if we define $\pi_z(\sigma)$ to be the product of terms $\lambda_z(\tau)$, one for each spanning subgraph of σ which has type τ, we see that π_z is the function satisfying the inverse relations (3.2.1) and (3.2.2). Since each type τ contributing to the product has the same rank as σ, the result (Theorem G) for π_z follows from the similar result for λ_z proved above.

4. PARTITION FUNCTIONS FOR INFINITE GRAPHS

4.1 Background

In this final section we shall examine one of the major problems associated with interaction models: the definition and properties of a partition function for infinite graphs. The original motivation came from theoretical physics. Physical systems have a large number of particles, and sometimes exhibit 'critical behaviour', such as phase transitions. It might be hoped that such phenomena will manifest themselves as singularities in the infinite limit.

Let us suppose that $\{G_n\}$ is a sequence of finite graphs which, in some sense, tends to the limit G_∞ (an infinite graph). If G_n has v_n vertices, then it is appropriate to consider the limit

$$\lim_{n\to\infty} \{Z(M,G_n)\}^{1/v_n} .$$

For example, in his original paper on the Ising model, Ising (1925) showed that for the sequence $\{C_n\}$ of circuit graphs,

$$Z(I_T, C_n) = [2\cosh(L/kT)]^n + [2\sinh(L/kT)]^n ,$$

and so in this case the limit exists and is $2\cosh(L/kT)$. If we consider that the infinite linear chain C_∞ as the limit of $\{C_n\}$ we might be tempted to write

$$Z(I_T, C_\infty) = 2\cosh(L/kT) \quad .$$

Clearly, there are very substantial problems here, concerning the existence and uniqueness of the limit. In general, we may distinguish three main techniques which have been applied to such problems: exact calculations, approximations, and 'theory'. We shall review them in turn.

4.2 Exact Calculations

For any sequence $\{G_n\}$ with the property that G_n has n-fold cyclic symmetry, it is possible that the 'transfer matrix' method [IM, p.26] will lead to exact results. Ising's calculation for the circuit graphs can be done in this way (although he used another method). The Ising result is satisfactory from one point of view, since the limit of $Z(I_T, C_n)^{1/n}$ exists for all temperatures $T > 0$, and it is a smooth function of T. However, from the physicists' viewpoint, the result is disappointing. The Ising model is intended to describe the behaviour of real magnetic materials, and such materials exhibit an abrupt change of behaviour at a certain critical temperature T_c.

The situation was rescued by Onsager (1944). He applied the transfer matrix method (and much else besides) to the study of the Ising model on the toroidal square lattice graphs $\square_n$ and their 'limit', the plane square lattice graph $\square_\infty$. He obtained a result involving a certain definite integral, which exists and is a smooth function of T for all positive values except one critical value, T_c. This remarkable result still casts its shadow over the whole field.

Since Onsager's time, the transfer matrix method has had a

few other notable successes. Lieb (1967) showed that

$$\lim_{n\to\infty} Z(C_3, \square_n)^{1/n^2} = (4/3)^{3/2} ,$$

but no exact results for the colouring model C_u on the plane square lattice with $u > 3$ are known. (Some years ago, I used the transfer matrix method to obtain approximations for $u > 3$ (Biggs, 1977b).)

4.3 Approximations

Most approximate methods are based on expansions similar to (2.4.2). Typically, we try to express the partition function $Z(M,G)$ as a sum of terms, one for each subgraph of G. If matters are arranged so that the smallest subgraphs (edges, triangles, and so on) give the largest contributions, then we can estimate $Z(M,G)$ by taking the first few terms of the sum. In the case of an infinite graph we shall obtain an infinite series, which might converge. Sadly, almost nothing has been proved rigorously about the behaviour of such series.

One possible approach is based on the methods outlined in Section 3. If we take the logarithm of the multiplicative formula (Theorem F), we obtain a series expansion for $\log \bar{Z}(R,G)$, to which only the non-separable subgraphs contribute. (This is often called a *cluster expansion*.) Furthermore, the contributions obey the law of vanishing coefficients (Theorem G).

Despite the notable absence of theoretical justification, the blind faith of the physicists has led them to draw many interesting conclusions from the study of such series (Domb, 1974). For example, it is known that the Ising model on the three-dimensional cubic lattice has a critical temperature T_c, although its exact value has not been calculated. The series expansion method enables us to estimate T_c with a high degree of accuracy, even though the series involved may not be convergent.

Series methods have been applied to the colouring model on $\square_\infty$ by Nagle (1971), Baker (1971), and Kim & Enting (1979). The last-named authors find the following power series in $x = 1/(u-1)$ for the limit of $\bar{Z}(C_u, \square_n)^{1/n^2}$:

$$(x+1)^{-2}[1 + x^3 + x^7 + 3x^8 + 4x^9 + 3x^{10} + 3x^{11} + 11x^{12} + 24x^{13} + 8x^{14} - 91x^{15} - 261x^{16} - 290x^{17} + 254x^{18} + \ldots] .$$

But we have no estimate of the radius of convergence of the series, and the behaviour of the signs of the coefficients is rather alarming.

4.4 Theory

It would be satisfying to have a procedure which leads to results of the following kind: for a given infinite graph G_∞, and any sequence $\{G_n\}$ of finite graphs which tends to G_∞ (in a sense to be defined), the limit of $Z(M, G_n)^{1/v_n}$ exists and is independent of the sequence $\{G_n\}$. Such theorems have been proved for particular graphs (such as $\square_\infty$) and particular interaction models (such as I_T). A typical formulation is given by Ruelle (1969). A more general approach may be based on a notion of dimensionality. Roughly speaking, an infinite graph is *d-dimensional* if (i) it admits the product Z^d of d infinite cyclic groups acting as a fixed-point-free group of automorphisms, and (ii) the number of orbits of Z^d is finite. Using this definition, Grimmett (1978a, 1978b) has established a strong result concerning the existence of the limit for certain sequences of subgraphs of a d-dimensional graph, and any interaction model. His proof makes use of a sub-additivity property and a related ergodic theorem. More could be proved about the uniqueness of the limit if a couple of reasonable conjectures were verified.

It appears that this field is worthy of further study. Thus far, the outstanding achievements have been the result of remarkably intricate calculations, directed towards problems of special interest to physicists. The mathematical foundations have yet to be established.

Royal Holloway College, University of London
Egham, Surrey TW20 OEX, England

REFERENCES

Baker, G.A. (1971). Linked-cluster expansion for the graph-vertex coloration problem. *J. Combinatorial Theory* (B), 10, 217-31.

Biggs, N.L. (1974). *Algebraic Graph Theory*. Cambridge University Press.

Biggs, N.L. (1977a). *Interaction Models*. L.M.S. Lecture Note Series 30, Cambridge University Press.

Biggs, N.L. (1977b). Colouring square lattice graphs. *Bull. London Math.Soc.* 9, 54-6.

Domb, C. (1974). Graph theory and embeddings. In *Phase Transitions and Critical Phenomena*, ed. C. Domb and M.S. Green, vol.3, pp.1-96. London: Academic Press.

Grimmett, G.R. (1978a). Multidimensional lattices and their partition functions. *Quart.J.Math. Oxford* (2), 29, 141-57.

Grimmett, G.R. (1978b). The rank polynomials of large random lattices. *J. London Math.Soc.* 18, 567-75.

Ising, E. (1925). Beitrag zur theorie des ferromagnetismus. *Z.Physik* 31, 253-8.

Kim, D. and Enting, I.G. (1979). The limit of chromatic polynomials. *J. Combinatorial Theory* (B) (to appear).

Lieb, E.H. (1967). The residual entropy of square ice. *Phys. Rev.* 162, 162-72.

Nagle, J.F. (1971). A new subgraph expansion for obtaining coloring polynomials for graphs. *J. Combinatorial Theory* (B), 10, 42-59.

Onsager, L. (1944). Crystal statistics, I. A two-dimensional model with an order-disorder transition. *Phys.Rev.* 65, 117-49.

Ruelle, D. (1969). *Statistical Mechanics.* Reading, Mass.: Benjamin.

Tutte, W.T. (1967). On dichromatic polynomials. *J. Combinatorial Theory* (B), 2, 301-20.

Tutte, W.T. (1979). All the King's Horses. In *Graph Theory and Related Topics*, ed. J.A. Bondy and U.S.R. Murty. New York: Academic Press.

Vout, C.W. (1978). *Functions on Graphs and some Generalizations.* Thesis, University of London.

Whitney, H. (1932). The colouring of graphs. *Ann.Math.* 3, 688-718.

2 · Symmetry conditions in graphs

A. Gardiner

INTRODUCTION

It is only too easy for the proliferation of results involving symmetry conditions in graphs to conform to no very obvious pattern. I should like to try to impose a simple conceptual framework on those sections of the subject in which I have been chiefly involved, by considering a sequence of examples which seem to me to emerge in a natural way. The treatment is by no means either uniform or exhaustive - emphasis and detail reflecting my own personal interests.

All graphs Γ will be finite, undirected, without loops or multiple edges, and are generally assumed to be connected. (Though the ideas usually extend in one way or another to directed graphs, the results and methods of proof rarely do.)

The framework I shall propose arises by first exploring the most extreme symmetry condition one might use to define a class of graphs; afterwards we consider the ways in which such extremism might be most profitably modified.

EXAMPLE I

A graph Γ is (globally) *homogeneous* if whenever $U,V \subseteq V\Gamma$ gives rise to isomorphic induced subgraphs $\langle U\rangle,\langle V\rangle$, every isomorphism $\sigma: \langle U\rangle \rightarrow \langle V\rangle$ extends to an isomorphism of Γ .

Here no restriction is placed on the isomorphism type of the induced subgraphs $\langle U\rangle,\langle V\rangle$, and no assumption is made about the way U and V are embedded in Γ . The assumption is distressingly strong, and one is not surprised to find a complete classification of such graphs.

Theorem (Sheehan, 1974; Gardiner, 1976a) Let Γ be a homogeneous graph. Then Γ is isomorphic to one of the following:

(i) the disjoint union $t.K_n$ $(t \geq 1)$ of t copies of the complete graph K_n ;

(ii) the circuit C_5 of length 5 ;

(iii) the regular complete t-partite graph $K_{t;r}$ with blocks of size r ;

(iv) the line graph $L(K_{3,3})$ of the complete bipartite graph $K_{3,3} \simeq K_{2;3}$.

The proof of the theorem is inductive: if Γ^o is such a graph, $u_o \in V\Gamma^o$, and $\Gamma^o(u_o)$ denotes the set of neighbours of u_o in Γ^o , then $\Gamma^1 = \langle\Gamma^o(u_o)\rangle$ is also such a graph; similarly, if $u_1 \in V\Gamma^1$, then $\langle\Gamma^1(u_1)\rangle$ is again such a graph. This process terminates when $\langle\Gamma^i(u_i)\rangle \simeq t.K_1$ is a null graph. Thus one need only show that the smallest class $\underline{\underline{X}}$ of homogeneous graphs which contains all null graphs, and which is 'closed with respect to extension' (that is, $\Gamma \in \underline{\underline{X}}$ whenever Γ is homogeneous and $\langle\Gamma(u)\rangle \in \underline{\underline{X}}$ for some $u \in V\Gamma$) is precisely the class of graphs listed in the theorem.

This approach is of little direct use for infinite graphs and in fact much more interesting things are possible for such graphs.

Even though the assumption that 'Γ is homogeneous' is in some sense the direct analogue of the basic axiom of motion geometry (given any two congruent configurations A,B in a space, then each congruence $\sigma: A \to B$ can be realised by a transformation of the whole space) it is evidently far too strong, and there are several distinct directions in which one might naturally try to relax it.

EXAMPLES IIA

The first generalisation might be motivated by observing that the motion axiom for congruent configurations in plane geometry follows from the motion axiom for triangles. It is thus natural to require homogeneity only for certain key isomorphism types of subgraphs: let $\underline{\underline{X}}$ be a class of graphs closed under isomorphism; a graph Γ is (globally) $\underline{\underline{X}}$-*homogeneous* if whenever $U,V \subseteq V\Gamma$ satisfy $\langle U\rangle \simeq \langle V\rangle \in \underline{\underline{X}}$, then each isomorphism $\sigma: \langle U\rangle \to \langle V\rangle$ extends to an automorphism of Γ.

By choosing $X = \{K_1\}$, $X = \{K_2\}$, etc., one obtains in this way the vertex-transitive, the symmetric, and many other familiar classes of graphs. But that was not the original aim - we wished rather to choose the class $\underline{\underline{X}}$ in such a way that the class of $\underline{\underline{X}}$-homogeneous graphs was both interesting and tractable. If we wish the original proof-strategy to apply, we shall seek to choose $\underline{\underline{X}}$ so that $\langle\Gamma(u)\rangle$ $(u \in V\Gamma)$ is $\underline{\underline{X}}$-homogeneous whenever Γ is; the simplest way to ensure this is to choose $\underline{\underline{X}}$ so that '$\Delta \in \underline{\underline{X}}$ implies $\Delta + K_1 \in \underline{\underline{X}}$' (Weiss, 1977a; Gardiner, 1978).

A less artificial example which arose independently and which has since proved to be of some significance is the class of 's-arc-transitive' graphs. In seeking to classify the minimal trivalent graphs of given girth g, Tutte separated the cases of even girth $g = 2s-2$ and odd girth $g = 2s-1$, and was led to consider the different problems of classifying trivalent $\{P_s\}$-homogeneous graphs for each s, where P_s denotes the path of length s. Here he proved a surprisingly strong result.

<u>Theorem</u> (Tutte, 1947) The class of trivalent $\{P_s\}$-homogeneous graphs is non-empty if and only if $s \leq 5$.

In the same paper Tutte classified the 'minimal' trivalent

$\{P_s\}$-homogeneous graphs for each s, $2 \le s \le 5$ - that is, the graphs Γ with diameter $d = s-1$ (it was these examples which seemed to solve his original problem): if $s = 2$, then $\Gamma \simeq K_4$; if $s = 3$, then $\Gamma \simeq K_{3,3}$ or Petersen's graph O_3; if $s = 4$, then Γ is the incidence graph of points and lines in the 7 point projective plane $PG(2,2)$; if $s = 5$, then Γ is the incidence graph of points and totally isotropic lines in the projective space $PG(3,2)$ endowed with a symplectic form (Tutte's 8-cage).

The classification of trivalent $\{P_s\}$-homogeneous graphs $(1 \le s \le 5)$ is still not in a satisfactory state. If the automorphism group Aut Γ acts primitively on $V\Gamma$ (as in the minimal examples $s = 2$, $\Gamma \simeq K_4$ and $s = 3$, $\Gamma \simeq O_3$) then the classification is complete (Wong, 1967). But one sees already from the 'minimal' graphs that there exist natural and important examples which cannot be built up from primitive graphs. These other 'minimal' graphs are all *semi-primitive* (that is, they are bipartite with the property that the stabiliser in Aut Γ of each part acts primitively on that part), and Wong's work can also be used to completely classify semi-primitive trivalent graphs (Gardiner, unpublished).

But the restriction to trivalent graphs runs counter to our general strategy. We should be asking: What can one say, for a given value of s, about the class of $\{P_s\}$-homogeneous graphs? If $s = 0$ or $s = 1$ then we obtain all vertex-transitive and all symmetric graphs respectively; thus we assume $s \ge 2$. It would be nice if every $\{P_{s+1}\}$-homogeneous graph Γ was necessarily $\{P_s\}$-homogeneous, but this is not quite true:

is $\{P_4\}$- homogeneous, but not $\{P_3\}$-homogeneous. However (finite) trees which are $\{P_s\}$- homogeneous can be easily described so we may restrict our attention to $\{P_{s+1}\}$-homogeneous graphs Γ that contain a circuit, in which case every subgraph isomorphic to P_s can be extended to a subgraph isomorphic to P_{s+1} (in both directions) so Γ is necessarily $\{P_s\}$-homogeneous.

Every $\{P_s\}$-homogeneous graph Γ $(s \geq 2)$ which contains a circuit is therefore $\{P_o\}$-homogeneous (i.e. vertex-transitive) and $\{P_2\}$-homogeneous. Thus if $G = \text{Aut}\,\Gamma$, then the vertex stabilisers are all conjugate subgroups of G, and the stabiliser $G(v)$ of any given vertex v must act transitively on the subgraphs $\langle u,v,w\rangle$ of Γ isomorphic to P_2 and having midpoint v; that is, $G(v)$ must act transitively on ordered pairs (u,w) of vertices adjacent to v, or in other words the permutation group $G(v)^{\Gamma(v)}$ induced by $G(v)$ on the set $\Gamma(v)$ of vertices adjacent to v in Γ is doubly transitive. The kernel of the action of $G(v)$ on $\Gamma(v)$ is the normal subgroup $G_1(v)$ of $G(v)$ fixing v and each of its neighbours:

$$G_1(v) = \bigcap_{w \in \Gamma(v)} G(vw)$$

where $G(vw) = G(v) \cap G(w)$. Thus each stabiliser $G(v)$ has a normal subgroup $G_1(v)$ such that $G(v)/G_1(v) \simeq G(v)^{\Gamma(v)}$ is a doubly transitive permutation group. Now this is encouraging since it is very likely that all doubly transitive permutation groups are known; though the final classification still eludes us there are several conditions relevant to our present problem under which all such groups are known.

But even if the possibilities for $G(v)/G_1(v)$ are known, is this really a cause for rejoicing? After all this only tells us about the action of the vertex stabiliser $G(v)$ on the neighbours of v: it seems to say nothing at all about the action of $G(v)$ on vertices further away from v, or about the subgroup $G_1(v)$. Fortunately this statement is at best half true (and may eventually prove to be wholly false). One reason for optimism is exemplified by the simplest interesting geometrical examples of $\{P_2\}$-homogeneous graphs - namely the incidence graph Γ of points and hyperplanes in a desarguesian projective space: if $G = \mathrm{Aut}\,\Gamma$, H is a hyperplane and x a point in H, then the subgroup $G_1(xH) = G_1(x) \cap G_1(H)$ consists of all elations with centre x and axis H, and so is an (elementary abelian) p-group for some prime p. This is typical of the general case.

Theorem (Gardiner, 1973) If Γ is $\{P_o\}$-homogeneous and $G = \mathrm{Aut}\,\Gamma$ is such that $G(u)$ acts primitively on $\Gamma(u)$ for some vertex u, then for each $v \in \Gamma(u)$ the group $G_1(uv)$ is a p-group for some prime p; this is true in particular if Γ contains a circuit and is $\{P_2\}$-homogeneous.

Since $G_1(u)/G_1(uv) \simeq G_1(u).G_1(v)/G_1(v) \lhd G(uv)/G_1(v) \simeq G(uv)^{\Gamma(v)} \simeq G(uv)^{\Gamma(u)}$ we see that $G(u)^{\Gamma(u)}$ determines the possibilities for $G_1(u)/G_1(uv)$, so the possibilities for $G(u)$ would be determined if we could only control $G_1(uv)$ in some way. Now $G_1(uv) = G_1(u) \cap G_1(v) \lhd G(uv)$; if $G_1(uv)$ is the maximal normal p-subgroup of $G_1(v)$, then $G_1(uv) = 1$. Thus either

(i) $G_1(uv) = 1$ and the possibilities for $G(u)$ are determined by $G(u)^{\Gamma(u)}$, or

(ii) $G_1(uv) \lneq O_p(G_1(v))$, so $G(u)^{\Gamma(u)}$ is a doubly transitive group in which the stabiliser $G(uv)^{\Gamma(u)}$ of a point $v \in \Gamma(u)$ has a non-trivial normal p-subroup $O_p(G_1(v)).G_1(u)/G_1(u)$.

Doubly transitive groups satisfying the condition in (ii) above have been extensively studied: every known group of this kind has a normal subgroup which is either regular or is isomorphic to one of PSL(n,q), Sz(q), PSU(3,q), or R(q) (a group of Ree type). There are very good grounds for believing that this list is complete. Consideration of the various known possibilities for $G(u)^{\Gamma(u)}$ yields results which may be summed up in the following theorem.

<u>Theorem</u> (Gardiner, 1973, 1974a, 1974b; Weiss, 1977b, in press a, in press b, in press c, unpublished; Bürker and Knapp, 1976; Dempwolff, 1976a, 1976b) Let Γ be a $\{P_s\}$-homogeneous graph $(s \geq 2)$ containing a circuit; let $G = \mathrm{Aut}\,\Gamma$, $u \in V\Gamma$, $v \in \Gamma(u)$.

(i) If $G(u)^{\Gamma(u)} \triangleright PSL((2,p^r)$, then $s \leq 4$, or $s = 5$ and $p = 2$, or $s = 7$ and $p = 3$; $G_1(uv) = 1$ if $s = 2$ and $|G_1(uv)| = (p^r)^{s-3}$ for $s \geq 3$.

(ii) If $G(u)^{\Gamma(u)} \triangleright Sz(q)$, $PSU(3,q^2)$, or $R(q)$, then $G_1(uv) = 1$ and $s \leq 3$.

(iii) If $G(u)^{\Gamma(u)} \triangleright PSL(n,p^r)$ with $n \geq 3$, then $s \leq 3$; if $p \geq 5$ and $s = 3$, then $|G_1(uv)|$ divides $(p^r)^{(n-1)^2}$.

(iv) If $G(u)^{\Gamma(u)}$ contains a regular normal subgroup and $|\Gamma(u)| \geq 5$, then $G_1(uv) = 1$ and $s \leq 3$.

<u>Corollary</u> Let Γ be a $\{P_s\}$- homogeneous graph containing a circuit and having valency $k \geq 3$. If $G = \mathrm{Aut}\,\Gamma$ then one of the following holds:

(i) $G_1(uv) = 1$.

(ii) $s \leq 3$.

(iii) $s = 4$, $G(u)^{\Gamma(u)} \triangleright PSL(2,p^r)$, $k = p^r+1$, $|G_1(uv)| = p^r$.

(iv) $s = 5$, $G(u)^{\Gamma(u)} \triangleright PSL(2,2^r)$, $k = 2^r+1$, $|G_1(uv)| = 2^{2r}$.

(v) $s = 7$, $G(u)^{\Gamma(u)} \triangleright PSL(2,3^r)$, $k = 3^r+1$, $|G_1(uv)| = 3^{4r}$.

(vi) $G(u)^{\Gamma(u)}$ is an unknown doubly transitive group in which the stabiliser of a vertex $v \in \Gamma(u)$ has a non-trivial normal p-subgroup.

As yet it remains an open problem to prove that '$G_1(uv) = 1$ implies $s \leq 3$', but it now looks reasonable to hope to classify $\{P_s\}$-homogeneous graphs with $s \geq 4$.

EXAMPLE IIB

Another direction in which the original homogeneity assumption might be weakened is to assume merely that whenever $U,V \subseteq V\Gamma$ satisfy $\langle U\rangle \simeq \langle V\rangle$, then *at least one* isomorphism $\sigma: \langle U\rangle \to \langle V\rangle$ extends to an automorphism of Γ; graphs Γ satisfying this condition will be called *weakly homogeneous*. One might suspect that making this assumption for *all* subsets U,V of $V\Gamma$ was still far too strong, but the crude inductive proof which was used for homogeneous graphs is not quite sufficient as it stands: if Γ is weakly homogeneous and $u \in V\Gamma$, then there is no obvious reason why $\langle\Gamma(u)\rangle$ should be weakly homogeneous (Gardiner, 1976a). But by restricting attention to a nested sequence of cliques $X_1 \subseteq X_2 \subseteq \ldots$, $\langle X_i\rangle \simeq K_i$, and the corresponding sequence $\Gamma(X_1) \supseteq \Gamma(X_2) \supseteq \ldots$, where $\Gamma(X) = \bigcap_{x\in X} \Gamma(x)$, Ronse (Ronse, 1978) produces a slightly different argument which suffices. Enomoto (Enomoto, in press) classifies graphs Γ satisfying a weaker condition which appears to say nothing about automorphisms of Γ: he considers simply graphs Γ in which whenever $U,V \subseteq V\Gamma$ satisfy $\langle U\rangle \simeq \langle V\rangle$, then $\langle\Gamma(U)\rangle \simeq \langle\Gamma(V)\rangle$. Thus we obtain the result.

<u>Theorem</u> (Gardiner, 1976a; Ronse, 1978; Enomoto, in press) Every weakly homogeneous graph is homogeneous.

The inductive proofs used in problems such as this focus

attention once again on the general extension problem: Given a graph Δ (or a class of graphs $\underline{\underline{D}}$), find all graphs Γ of some specified kind such that $\langle\Gamma(u)\rangle \simeq \Delta$ (or $\langle\Gamma(u)\rangle \in \underline{\underline{D}}$) for each $u \in V\Gamma$ (see for example Hell, 1976; Hubaut, 1976).

EXAMPLES IIC

Yet another fruitful way of modifying the definition of homogeneity is to assume free mobility only between subgraphs which are not only isomorphic but are also in some sense 'similarly embedded' in Γ. It is not entirely clear how to specify this intuitive idea in practice, but we shall restrict our attention here to a single interpretation. The class $\underline{\underline{T}}_n$ consists of those connected graphs Γ such that whenever $(u_1,\ldots,u_n)$ and $(v_1,\ldots,v_n)$ are n-tuples of vertices of Γ satisfying $\partial(u_i,u_j) = \partial(v_i,v_j)$ for all i,j (where ∂ is the natural distance function), then there is an automorphism α of Γ such that $\alpha(u_i) = v_i$ for all i. Thus $\underline{\underline{T}}_2$ is the class of distance transitive graphs, certain aspects of which will be discussed below. $\underline{\underline{T}}_3$ consists of Meredith's triple-transitive graphs (Meredith, 1976); Meredith showed any such graph has valency $k \leq 2$ or girth $g = 3$ or 4, and that the number of such graphs of girth 4 and valency k was finite for each k. Cameron observed that if $\Gamma \in \underline{\underline{T}}_3$, $G = \mathrm{Aut}\,\Gamma$, and $u \in V\Gamma$, then $G(u)^{\Gamma(u)}$ is triply transitive and was then able to use earlier work (Cameron, 1972) to prove:

<u>Theorem</u> (Meredith, 1976; Cameron, 1977) Let $\Gamma \in \underline{\underline{T}}_3$ have girth $g \geq 4$ and valency k. Then one of the following occurs:

(i) $\Gamma \simeq K_{k,k}$;

(ii) Γ is obtained from $K_{k+1,k+1}$ by deleting the edges of a matching;

(iii) $\Gamma = \Gamma_H$, where H is a Hadamard matrix of Sylvester type $(k = 2^d)$ or of order 12 $(k = 12)$;

(iv) Γ has diameter 2 , and for some positive integer $\mu, k = (\mu+1)(\mu^2+5\mu+5)$, and $c_2 = (\mu+1)(\mu+2)$ is the number of paths of length 2 joining two vertices at distance 2 ;

(v) Γ is obtained from a graph of type (iv) by doubling;

(vi) $\Gamma \simeq Q_k$ (the k-dimensional cube);

(vii) $\Gamma \simeq \square_k$, the graph obtained from Q_k by identifying antipodal vertices;

(viii) $\Gamma \simeq C_n$ $(n \geq 4)$.

More recently, Cameron has proved rather more than this, and has in particular classified all graphs in $\underline{\underline{T}}_6$.

<u>Theorem</u> (Cameron, in press) If $\Gamma \in \underline{\underline{T}}_6$, then Γ is one of the following:

(i) $K_{t;r}$;

(ii) $K_{k+1,k+1}$ with the edges of a matching deleted;

(iii) C_n ;

(iv) $L(K_{3,3})$;

(v) the icosahedron;

(vi) the graph whose vertices are the 3-subsets of a 6-set, two vertices adjacent whenever their intersection is a 2-set.

In particular if $\Gamma \in \underline{\underline{T}}_6$, then $\Gamma \in \underline{\underline{T}}_n$ for all $n \geq 1$.

As these results suggest, the class $\underline{\underline{T}}_2$ of distance-transitive graphs is rather special, yet sufficiently rich to be interesting and we shall survey its treasures and frustrations in some detail.

We should admit at the outset that though the assumption that a graph Γ is distance-transitive can be used quite effectively in deciding its existence or non-existence once we know the valency k, the diameter d, the size $k_i = |\Gamma_i(u)|$ of the ith 'circle' $\Gamma_i(u) = \{v \in V\Gamma: \partial(u,v) = i\}$ for each i, and a bit more (Biggs and Smith, 1971; Gardiner, 1975; Biggs, 1976; Doro, unpublished; Gordon and Levingston, unpublished; Buekenhout and Rowlinson, unpublished) we know hardly any general theorems which successfully exploit the assumption '$\Gamma \in \underline{\underline{T}}_2$'. Partly for this reason one is led to study graphs which satisfy a somewhat weaker condition.

Lemma Let $\Gamma \in \underline{\underline{T}}_2$ and $u,v \in V\Gamma$ with $\partial(u,v) = i$. Then $|\Gamma_{i-1}(u) \cap \Gamma(u)| = c_i$, $|\Gamma_i(u) \cap \Gamma(v)| = a_i$, $|\Gamma_{i+1}(u) \cap \Gamma(v)| = b_i$ are functions only of i and do not depend on the choice of u,v.

Clearly $c_o = 0$ since $\Gamma_{-1}(u)$ is not defined, and $b_d = 0$ since $\Gamma_{d+1}(u) = \phi$. We are thus led to the following definitions: a graph Γ of diameter d is *distance-regular* if for each $i \leq d$ there exist integers c_i, a_i, b_i depending only on i, such that for each pair of vertices u,v with $\partial(u,v) = i$, $|\Gamma_{i-1}(u) \cap \Gamma(v)| = c_i$, $|\Gamma_i(u) \cap \Gamma(v)| = a_i$, $|\Gamma_{i+1}(u) \cap \Gamma(v)| = b_i$. Thus every graph $\Gamma \in \underline{\underline{T}}_2$ is distance-regular; on the other hand, the assumption that Γ is distance-regular says nothing at all about Aut Γ, so one one cannot expect such a graph to be distance-*transitive*. Nevertheless *many* (and for small values of k and d, *most*) distance-regular graphs are distance-transitive, so it makes sense as a reasonable first approximation to seek necessary conditions for the existence of distance-*regular* graphs. Now each distance-regular graph Γ is necessarily regular of valency $k = b_o$, and determines the $3(d+1)$ parameters c_i, a_i, b_i $(0 \leq i \leq d)$; since $c_o = 0 = b_d$, and since $c_i + a_i + b_i = b_o$, we may restrict attention to the $2d$ parameters $\{c_1, c_2, \ldots, c_d; b_o, b_1, \ldots, b_{d-1}\}$ which constitute the *parameter array* of Γ (in fact $c_1 = 1$, but it is convenient to include c_1 nevertheless). It is natural to frame any necessary conditions for the existence of distance-regular graphs in terms of this parameter array.

If Γ is a distance-regular graph, $u \in V\Gamma$, $0 < i < d$, and $w \in \Gamma_{i+1}(u)$, then $|\Gamma(u) \cap \Gamma_i(w)| = c_{i+1} \neq 0$, so we may choose $v \in \Gamma(u) \cap \Gamma_i(w)$. Then $\Gamma(w) \cap \Gamma_{i-1}(v) \subseteq \Gamma(w) \cap \Gamma_i(u)$ so $c_i = |\Gamma(w) \cap \Gamma_{i-1}(u)| \leq |\Gamma(w) \cap \Gamma_i(u)| = c_{i+1}$. Similarly one can show that the b_i are non-increasing. Other elementary (but sometimes complicated) arguments lead to various other necessary conditions which a parameter array must satisfy if it is to originate from a distance-regular graph.

<u>Theorem</u> Let Γ be a distance-regular graph with diameter array $\{c_1,\ldots,c_d; b_o,\ldots,b_{d-1}\}$. Then

(i) (Biggs, 1974) $1 = c_1 \leq c_2 \leq \ldots \leq c_d$; $k = b_o \geq b_1 \geq \ldots \geq b_{d-1}$;

(ii) (Gardiner, in press)

(a) if $d \geq i + j$, then $c_i \leq b_j$;

(b) if $d > i > j$ and $a_{j+1} = a_{j+2} = \ldots = a_i = 0$, then $a_o = a_1 = \ldots = a_{i-j} = 0$; if furthermore $a_{i+1} \neq 0$, then $c_{i-j+1} \leq a_{i+1} \leq b_{i-j}$;

(c) if $b_1 = b_2 = \ldots = b_{i+1}$, then $c_1 = c_2 = \ldots = c_{i+1}$;

(d) if $c_3 = c_2$, then $c_3 = c_2 = c_1$.

These conditions suggest that the parameter arrays of distance-regular graphs have a distinct 'shape' which might be described much more precisely.

The assumption that Γ is distance-regular has powerful algebraic consequences: for the remainder of section IIC we assume that Γ is a distance-regular graph and explore the known consequences. Let A_i denote the $n \times n$ matrix ($n = |V\Gamma|$) having a 1 in row u and column v when $\partial(u,v) = i$, and a 0 otherwise; then $A = A_1$ is the adjacency matrix of Γ, and the distance-regular condition states precisely that

$$A.A_i = b_{i-1}A_{i-1} + a_iA_i + c_{i+1}A_{i+1} \quad (0 \le i \le d) \quad .$$

Hence $c_2A_2 = A_2 - a_1A - b_oI$, so A_2 is a polynomial $v_2(A)$ of degree 2 in A, $c_3A_3 = A.A_2 - a_2A_2 - b_1A$, so A_3 is a polynomial $v_3(A)$ of degree 3 in A, and so on. Thus A_d is a polynomial $v_d(A)$ of degree d in A, and $A.A_d = b_{d-1}A_{d-1} + a_dA_d$ yields a polynomial of degree $d+1$ satisfied by A; since a graph of diameter d has at least $d+1$ distinct eigenvalues, this must be the minimal polynomial of A. Thus the powers of A generate a $d+1$ dimensional algebra $\underline{\underline{A}}$ of $n \times n$ matrices with bases $\{I, A, \ldots, A^d\}$ and $\{A_o = I, A_1, \ldots, A_d\}$. All this has two important consequences:

(i) $A_hA_i \in \underline{\underline{A}}$, so $A_hA_i = \sum_{j=0}^{d} s_{hij}A_j$ for some s_{hij} (which must be non-negative integers, since in combinatorial language the equation states that if $u,v \in V\Gamma$, then $|\Gamma_h(u) \cap \Gamma_i(v)| = s_{hij}$ depends only on $j = \partial(u,v)$). Note that $s_{1i\,i-1} = b_{i-1}$, $s_{1ii} = a_i$, $s_{1i\,i+1} = c_{i+1}$.

(ii) (a) If we represent $\underline{\underline{A}}$ as an algebra of linear transformations of itself (via left multiplication) then we obtain an isomorphism of $\underline{\underline{A}}$ to an algebra of $d+1 \times d+1$ matrices which as we see from (i), maps A_h to $B_h = (s_{hij})_{ji}$. Thus the eigenvalues $\lambda_o = k, \lambda_1, \ldots, \lambda_d$ of A are precisely the eigenvalues of

$$B = B_1 = \begin{bmatrix} 0 & c_1 & & & \\ k & a_1 & c_2 & & \\ & b_1 & a_2 & \cdot & \\ & & b_2 & \cdot & c_d \\ & & & \cdot & a_d \end{bmatrix}$$

(b) The multiplicity $m(\lambda_i) = n/(\sum_{j=0}^{d} v_j(\lambda_i)^2/k_j)$ of λ_i as an eigenvalue of A can be computed from B and must be an integer. Moreover since the correspondence $A_h \leftrightarrow B_h$ induces an algebra isomorphism we must have $B.B_h = b_{h-1}B_{h-1} + a_hB_h + c_{h+1}B_{h+1}$ $(0 \le h \le d)$ and hence $B_h = v_h(B)$ is a polynomial of degree h in B having non-negative integral entries.

In certain cases one can apply these necessary conditions, which B (and hence the parameter array $\{c_1,\ldots,c_d; b_0,\ldots, b_{d-1}\}$) must satisfy if they are to originate from a distance-regular graph, directly and effectively to study infinite families of parameter arrays. The best known results of this type are perhaps (Feit and Higman, 1964; Singleton, 1966; Bannai and Ito, 1973; Damerell, 1973). But in many cases it is impossible to handle a given infinite family of parameter arrays in a uniform way, and one must first use some extraneous method to restrict to a finite number of cases before applying our necessary conditions to the remaining possibilities. And it is usually in this rather unsatisfactory way that we find ourselves classifying distance-transitive graphs - the extraneous results that we use here are about permutation groups. Suppose for example that we wish to classify distance-transitive graphs Γ of valency 3 (resp. valency $p+1$, p an odd prime). Then we have a group $G = \mathrm{Aut}\,\Gamma$ of automorphisms of a $\{P_1\}$- homogeneous graph Γ, so by Tutte's theorem (Tutte, 1947) (resp. by Gardiner, 1973, 1976b) we know that $|G(u)| = 3.2^{s-1}$ for some s, $1 \le s \le 5$ (resp. $|G(u)|$ divides $p+1.p^{s-1}.p-1$ for some s, $1 \le s \le 4$, or $p = 3$ and $s = 7$). But $G(u)$ must act transitively on each circle $\Gamma_i(u)$, so $k_i = |\Gamma_i(u)| = b_0b_1\ldots b_{i-1}/c_2c_3\ldots c_i$ divides $|G(u)|$. But the monotonicity of the b_i and the c_i implies that the k_i are 'unimodal' (that is, increase strictly, remain constant for a while, and may then decrease strictly). Thus if we can show that at most $j+1$ of the

k_i's can be equal, then we have only a finite number of parameter arrays to consider. Two very easy lemmas (Biggs and Smith, 1971) show that in the trivalent case $j \leq 2$, so the worst conceivable sequence of k_i's would be 1,3,6,12,24, 48,48,48,24,12,6,3,1. Thus we may then proceed either by programming a machine to list only those arrays with $d \leq 12$ and $k = 3$ which satisfy our necessary conditions (Biggs and Smith, 1971), or by translating the distance-transitive condition into permutation group theoretic language in the hope of understanding why (in some sense) so few examples exist (Gardiner, 1975).

<u>Theorem</u> (Biggs and Smith, 1971) There are just twelve trivalent distance-transitive graphs.

Both approaches extend to the case $k = 4$ (Smith, 1973, 1974a, 1974b; Gardiner, unpublished).

<u>Theorem</u> There are just fifteen tetravalent distance-transitive graphs.

The first strategy has been extended (Biggs, 1976) to list all parameter arrays with $3 \leq k \leq 13$, $2 \leq d \leq 5$ which satisfy the necessary conditions mentioned above (and one further condition - the so-called Krein condition); but before we discuss Biggs' list we must digress slightly.

Distance-transitive graphs exhibit a striking compatibility between their combinatorial (graph) structure and the properties of the permutation group $G = \text{Aut}\,\Gamma$, which is exceedingly useful in practice. As a permutation group on the vertex set $V\Gamma$, G may be either primitive or imprimitive. But there are two rather obvious ways in which a distance-transitive graph Γ may admit a system of imprimitivity: (i) Γ may be *bipartite* (like the cube Q_3) in which case the two halves of the bipartition form a system of imprimitivity, or (ii) Γ may be *antipodal* (like the icosahedron, or the cube) in the sense that $\{\{u\} \cup \Gamma_d(u): u \in V\Gamma\}$ is a partition of $V\Gamma$, whose parts are called the *antipodal blocks* of Γ. In fact these are the only possibilities.

<u>Theorem</u> (Smith, 1971) If Γ is a distance-transitive graph and $G = \text{Aut}\,\Gamma$ acts imprimitively on $V\Gamma$, then Γ is antipodal or bipartite.

In practice the natural assumption about the action of G on $V\Gamma$ seems to be *not* that it is primitive, but that it is *semi-primitive* (that is, $G^+ = \langle G(u): u \in V\Gamma \rangle$ acts primitively on each of its orbits).

The combinatorial notions of 'bipartite' and 'antipodal' apply equally well to distance-regular graphs, and Smith's result extends in a natural way to distance-regular graphs (Gardiner, unpublished). The notion of 'semi-primitive' also extends (a distance-regular graph Γ is *semi-primitive* if it is either complete bipartite or not antipodal of even diameter).

If Γ is an antipodal distance-regular graph, then the antipodal blocks form the vertices of a new graph $\bar{\Gamma}$, which is distance-regular of valency $\bar{k}$ (and $\bar{k} = k$ unless $\Gamma \simeq K_{t;r}$). Γ is then called a $(1+k_d)$-fold *antipodal covering* of $\bar{\Gamma}$, and $\bar{\Gamma}$ is the *antipodal quotient* of Γ. This idea is important for two reasons: (i) $\bar{\Gamma}$ is necessarily semi-primitive, and (ii) $1+k_d \le k$. It is thus natural to split the classification of distance-regular, or of distance-transitive, graphs of valency k into two parts: (i) classify the semi-primitive graphs of valency k, and (ii) construct all r-fold antipodal coverings of semi-primitive graphs $(2 \le r \le k)$.

Given the parameter array of a semi-primitive graph one

can construct the parameter arrays, and hence the spectra, of all possible r-fold antipodal coverings of Γ (if any such are possible), and can then apply to these possible arrays the standard necessary conditions discussed above. In this way some general covering problems can be satisfactorily resolved (for example,

<u>Theorem</u> (Gardiner, 1974c) A distance-regular graph of valency k admits a k-fold antipodal covering if and only if $\Gamma \simeq C_n, K_6, K_{56}$ (and a Moore graph of valency 57 exists), Tutte's trivalent 8-cage, or $K_{k,k}$ (and a projective plane of order k exists));

whereas at the other extreme some covering problems are really very hard (for example, 2-fold antipodal coverings of complete graphs K_n correspond to regular 2-graphs on n points whose classification is complete only for $n \leq 28$, and whereas there are only 10 non isomorphic non-trivial examples with $n \leq 28$, there are 91 known non-trivial examples with $n = 36$ (Seidel and Taylor, in press).

As well as being able to construct antipodal quotients of antipodal graphs, there is also a construction (Smith, 1971) which decomposes a bipartite semi-primitive distance-regular graph Γ into two 'primitive halves', which are cospectral though not necessarily isomorphic, and whose spectrum can be computed from that of Γ. Conversely given a distance-regular graph Δ one can derive necessary conditions which the parameter array of Δ must satisfy if it can be doubled in this way (Biggs and Gardiner, unpublished).

It is in this spirit that Biggs restricts his attention to those arrays with $2 \leq d \leq 5$, $3 \leq k \leq 13$ which might arise from *primitive* distance-transitive graphs (Biggs, 1976). The case $d = 2$, $k \leq 13$ had already been thoroughly discussed (Biggs, 1971). Of the remaining arrays, thirteen corresponded to graphs in recognised infinite families. Biggs was left with seven arrays realised by sporadic examples, and eleven undecided arrays. The combined and overlapping efforts of at least three sets of authors (Doro, unpublished; Buekenhout and Rowlinson, unpublished; Gordon and Levingston, unpublished) still do not quite complete the picture: three of the eleven cannot be realised by a distance-transitive graph, six can definitely be so realised (at least four of them uniquely) and the remaining two ($\{10,8,8,2; 1,1,4,5\}$ and $\{12,8,6,4; 1,1,2,9\}$) seem to be not completely resolved.

EXAMPLES III

Finally we consider conditions of a local nature. We shall be very brief even though it is arguable that certain conditions of this kind are more natural, more challenging, and potentially more fruitful than any of those to which so much space has been devoted above.

There are at least two distinct interpretations of the adjective 'local'. Firstly we might drop the idea of mobility *between* isomorphic, or similarly embedded, subgraphs and define Γ to be *locally* $\underline{\underline{X}}$*-homogeneous* if whenever $U \subseteq V\Gamma$ and $\langle U \rangle \in \underline{\underline{X}}$, each automorphism of $\langle U \rangle$ extends to an automorphism of Γ (Gardiner, 1978). The challenge here is to identify the class $\underline{\underline{X}}$ for which this local mobility of subgraphs in $\underline{\underline{X}}$ controls the structure of Γ in some desired way. A second possibility, and the one on which we shall focus our attention for the rest of this section, is to define Γ to be *locally* $\underline{\underline{X}}$*-homogeneous* if whenever $U, V \subseteq V\Gamma$ satisfy $\langle U \rangle \simeq \langle V \rangle \in \underline{\underline{X}}$, $w \in U \cap V$, and there exists an isomorphism $\sigma : \langle U \rangle \to \langle V \rangle$ fixing w, then α extends to an automorphism of Γ. The most promising work known to me is that on locally $\{P_s\}$-homogeneous graphs, the important novelty being that such graphs are not necessarily vertex-transitive, and may indeed contain vertices of two entirely different kinds; this opens up a completely new world, and provides a tool for studying both the incidence graphs of points and blocks in block designs and the intersection graphs of cosets corresponding to distinct subgroups of an abstract group.

Locally $\{P_s\}$-homogeneous graphs Γ in which the stabiliser in $\mathrm{Aut}\,\Gamma$ of each path of length s is trivial were introduced as a possible generalisation of Tutte's work on trivalent graphs (Bouwer and Djoković, 1973), but for interesting examples with valency $k > 3$ stabilisers of s-arcs are definitely non-trivial. Weiss (Weiss, 1976) considered locally $\{P_s\}$-homogeneous graphs $\Gamma, G = \mathrm{Aut}\,\Gamma$, with the property that for each vertex u, $G(u)$ does not act transitively on paths of length $s+1$ starting at u. This condition, though still not entirely satisfactory, is realised in many geometric examples, and Weiss extended many of the results on '$\{P_s\}$-homogeneous graphs which are not $\{P_{s+1}\}$-homogeneous' to this case (Weiss, in press a, in press d); in particular he discusses in detail the interesting case $G(u)^{\Gamma(u)} \trianglerighteq PSL(2,q)$. More recently David Goldschmidt became interested, for purely group-theoretical reasons, in the general problem of locally $\{P_s\}$-homogeneous graphs; in the trivalent case he obtained the bound $s \le 8$, and has since extended this result (Goldschmidt, unpublished). Rather surprisingly Goldschmidt's proof switches attention from the graph Γ which is of direct interest to an infinite regular trivalent tree, a switch which has been advocated for slightly different reasons by Djoković (Djoković, in press).

Pure Mathematics, University of Birmingham, England

REFERENCES

Bannai, E. and Ito, S. (1973). On Moore graphs. *J.Fac.Science University of Tokyo* IA, 20, 191-208.

Biggs, N.L. (1971). Finite groups of automorphisms. *London Math.Soc. Lecture Notes*, No.6, Cambridge Univ. Press, London.

Biggs, N.L. (1974). Algebraic graph theory. *Cambridge Math. Tracts*, No.67, Cambridge Univ. Press, London.

Biggs, N.L. (1976). Automorphic graphs and the Krein condition. *Geometriae Dedicata* 5, 117-27.

Biggs, N.L. and Smith, D.H. (1971). On trivalent graphs. *Bull. London Math.Soc.* 3, 155-8.

Bouwer, I.Z. and Djoković, D.Ž. (1978). On regular graphs III. *J. Combinatorial Theory* Ser.B, 14, 268-77.

Bürker, M. and Knapp, W. (1976). Zur Vermutung von Sims über primitive Permutationsgruppen II. *Arch.Math.* 27, 352-9.

Cameron, P.J. (1972). Permutation groups with multiply transitive suborbits. *Proc. London Math.Soc.* III Ser. 25, 427-40.

Cameron, P.J. (1977). A note on triple transitive graphs. *J. London Math.Soc.* (2) 15, 197-8.

Cameron, P.J. (in press). 6-transitive graphs. *J. Combinatorial Theory*.

Damerell, R.M. (1973). On Moore graphs. *Proc. Cambridge Philos.Soc.* 74, 227-36.

Dempwolff, U. (1976a). A factorisation lemma and an application. *Arch.Math.* 27, 18-21.

Dempwolff, U. (1976b). A factorisation lemma and an application II. *Arch.Math.* 27, 476-9.

Djoković, D.Ž. (in press). Automorphism of regular graphs and finite simple group-amalgams. *Proc.Intern.Colloqu. Algebraic methods in graph theory*. Szeged 1978.

Enomoto, H. (in press). Combinatorially homogeneous graphs. *J. Combinatorial Theory* Ser.B.

Feit, W. and Higman, G. (1964). The non existence of certain generalised polygons. *J. Algebra* 1, 114-31.

Gardiner, A. (1973). Arc transitivity in graphs. *Quart.J.Math. Oxford* (2), 24, 399-407.

Gardiner, A. (1974a). Arc transitivity in graphs II. *Quart.J. Math. Oxford* (2), 25, 163-7.

Gardiner, A. (1974b). Doubly primitive vertex stabilisers in graphs. *Math.Z.* 135, 257-66.

Gardiner, A. (1974c). Antipodal covering graphs. *J. Combinatorial Theory* Ser.B, 16, 255-73.

Gardiner, A. (1975). On trivalent graphs. *J. London Math.Soc.* (2), 10, 507-12.

Gardiner, A. (1976a). Homogeneous graphs. *J. Combinatorial Theory* Ser.B, 20, 94-102.

Gardiner, A. (1976b). Arc transitivity in graphs III. *Quart. J.Math. Oxford* (2), 27, 313-23.

Gardiner, A. (1978). Homogeneity conditions in graphs. *J. Combinatorial Theory* Ser.B, 24, 301-10.

Gardiner, A. (in press). When is an array realised by a distance-regular graph? *Proc.Intern.Colloqu. Algebraic methods in graph theory*. Szeged 1978.

Hell, P. (1976). Graphs with given neighbourhoods I. *Colloques Internationaux CNRS Problèmes Combinatoires et Théorie des Graphes*. Orsay 1976.

Hubaut, X. (1976). Extensions de Graphes. *Colloques Internationaux CNRS Problèmes Combinatoires et Théorie des Graphes*. Orsay 1976.

Meredith, G.H.J. (1976). Triple transitive graphs. *J. London Math.Soc.* (2), 13, 249-57.

Ronse, C. (1978). On homogeneous graphs. *J. London Math.Soc.* (2), 17, 375-9.

Seidel, J.J. and Taylor, D.E. (in press). Two-graphs, a second survey. *Proc.Intern.Colloqu. Algebraic methods in graph theory*. Szeged 1978.

Sheehan, J. (1974). Smoothly embeddable subgraphs. *J. London Math.Soc.* (2), 9, 212-18.

Singleton, R.R. (1966). On minimal graphs of maximal even girth. *J. Combinatorial Theory* 1, 306-32.

Smith, D.H. (1971). Primitive and imprimitive graphs. *Quart. J.Math. Oxford* (2), 22, 551-7.

Smith, D.H. (1973). On tetravalent graphs. *J. London Math. Soc.* (2), 6, 659-62.

Smith, D.H. (1974a). Distance-transitive graphs of valency four. *J. London Math.Soc.* (2), 8, 377-84.

Smith, D.H. (1974b). On bipartite tetravalent graphs. *Discrete Math.* 10, 167-72.

Tutte, W.T. (1947). On a family of cubical graphs. *Proc. Cambridge Philos.Soc.* 43, 459-74.

Weiss, R. (1976). Über lokal s-reguläre Graphen. *J. Combinatorial Theory* Ser.B, 20, 124-7.

Weiss, R. (1977a). Glatt einbettbare Untergraphen. *J. Combinatorial Theory* Ser.B.

Weiss, R. (1977b). Über symmetrischen Graphen und die projektiven Gruppen. *Arch.Math.* 28, 110-12.

Weiss, R. (in press a). Groups with a (B,N)-pair and locally transitive graphs. *Nagoya Math.J.*

Weiss, R. (in press b). An application of p-factorisation methods to symmetric graphs. *Proc. Cambridge Philos.Soc.* 85, 43-8.

Weiss, R. (in press c). Symmetric graphs with projective subconstituents. *Proc.Amer.Math.Soc.*

Weiss, R. (in press d). Elations of graphs. *Acta Math. Acad. Sci. Hungar.*

Wong, W.J. (1967). Determination of a class of primitive permutation groups. *Math.Z.* 99, 235-46.

3 · Extremal hypergraph problems

D.J. Kleitman

This talk is a review of recent developments in the study of extremal properties of set systems or hypergraphs.

We will begin with some examples of classical problems in this area; we will then discuss some recent results, and will finally discuss several conjectures and open problems. We apologize in advance for slighting important work (or contributors); in defence we note that while a review article can try to be encyclopaedic, nobody should be asked to listen to an encyclopaedic talk.

A graph is generally defined as a set of vertices and a set of pairs of vertices called arcs or lines or edges. A hypergraph similarly has vertices and hyperedges, which latter are here sets of vertices. A hypergraph is therefore another name for a collection of subsets of the vertices. We will refer particularly to several kinds of hypergraphs. A *k-hypergraph* has each of its members consisting of k vertices; an *intersecting hypergraph* has every pair of edges having at least one vertex in common. An *antichain* has no edge containing another; and an *ideal* has the property that any set of vertices contained in an edge is also an edge.

An *extremal property* is a statement that whatever possesses that property has to be or cannot be bigger than something or other.

The area that we are concerned with is not really extremal properties of hypergraphs per se (these are staggeringly trivial) but rather extremal properties of kinds of hypergraphs. Of course almost every mathematical construction contains within it some sort of hypergraph - so that almost any extremal property relates somehow to a kind of hypergraph. Our subject matter therefore really consists of extremal problems among those classes of hypergraphs that are not singled out by other more common names.

Let us look at some typical questions. More than fifty years ago Sperner answered the question, how large can an antichain on n vertices be? Erdös, Ko and Rado, a long time ago gave an upper bound on the size of an intersecting k-hypergraph on n vertices.

An old conjecture of Kneser (recently proven) suggests how many intersecting k-hypergraphs on n vertices are needed if their union is to contain all the k element sets of vertices.

Many questions have been raised, of the form: given a class of hypergraphs whose members obey restrictions on the size of their edges, of intersections and/or unions of edges, how large is their largest member? And there are many variations of this form which have come up.

There are also many questions having quite a different structure. Here for example is Chvátal's conjecture: given an ideal I, there is no intersecting hypergraph whose intersection with I is strictly larger than the number of edges of I containing some one vertex.

We give just two more examples here of the structures of an extremal question: How many vertices can we have so that there is a k-hypergraph A whose induced k-hypergraph on no set of j vertices is either complete or empty? This is a Ramsey type question. (By the induced hypergraph on some j vertices we mean the hypergraph on those vertices whose edges are the edges of A containing these vertices only.) Given two hypergraphs, another question is: What can one say about the hypergraphs consisting of unions and intersections of their edges? We will say more about this in (III.4).

We will now try to give an impression of the classical results of this area.

Before doing so we consider another question. Why and how do questions of this kind get asked? A moment's reflection tells us that any question of the size of anything probably has within it a question about the size of hypergraphs of some sort, since collections of sets are fundamental and general constructions. What is surprising here is that when all the details and structure of mathematical entities are stripped away and constructions are considered only as collections of sets that there should be enough information left so that answers to our hypergraphic questions say anything non-trivial about some original interesting question.

Strangely enough this sometimes happens. And this is in part due to the fact that we can give pretty sharp answers to some of these hypergraphic extremal problems.

II. A CLASSICAL RESULT

The task of this section is to illustrate the kind of result and kind of method that exists in this area. Now in a sense this task is absurd - no mathematical subject that could be adequately summed up by one result could be worth the effort. We mean thereby to give an illustration not a summary.

We begin by addressing the following question: we are to construct a hypergraph H on n vertices having E edges. The boundary $B(H)$ contains all edges not in H that differ from members of H in only one vertex. (A differs from C in only one vertex if $A \supset C$ or $C \subset A$, and $||A| - |C|| = 1$.) How can we do it so that $B(H)$ has minimum size? This "isoperimetric problem" has a very simple answer that, when looked at in the right way implies Sperner's Theorem, Erdös-Ko-Rado's theorem and all sorts of refinements of such questions. We will state the result, then outline a proof, and then follow some of its implications.

What is the answer? Well, there is obviously no unique

answer. Given some H we can permute the vertices or take a given vertex v out of all edges in H it was in, and put it into all edges of H it was not in and get another H without changing the size of $B(H)$; but we can give the following answer.

We order the vertices from 1 to n and refer to an edge by a sequence of zeros and ones, the i-th entry of which is one when the i-th vertex is in the edge. We add a zeroth entry so that the sum of all the entries in the sequence is n for each edge.

We associate an $n+1$ digit number $\theta(e)$ with a sequence or edge e by choosing the left-most digit to be the o-th entry, next the 1st ... and the right-most digit the n-th.

Then H will minimize the size of $B(H)$ if it consists of the edges or subsets or sequences having the E largest θ numbers.

What does this mean? Well, it says, first choose the empty edge, then the 1-edges, 2-edges, etc.; for any k , in choosing the k-edges choose the ones that are largest as numbers first, stop when you have E edges, and you minimize $B(H)$.

There are a number of simple proofs of this result, and some generalize to prove the same sort of result for hypergraphs of multisets, etc.

Here is one of the easier and most powerful proofs. It proceeds by induction on n ; the induction hypothesis is the statement of the theorem itself, which we may as well give here:

<u>Theorem</u> A hypergraph on n vertices having E edges that minimize its boundary can be formed by choosing the E edges having largest θ values.

We define an operation called *fixing*. Operating on a

hypergraph Z having E edges, Z_j of which contain the j-th vertex it creates a hypergraph $f_j(Z)$ which consists of the Z_j edges having largest θ values among those that contain the j-th vertex, and the $E-Z_j$ having largest θ values among the rest. Now the following facts are easy to show. (Nobody should try to show such things in a lecture. Everyone should try to do them at home.)

$$|B(f_j(Z))| \le |B(Z)| \tag{1}$$

$$\sum_{e \in f_j(Z)} \theta(e) \ge \sum_{e \in Z} \theta(e) \tag{2}$$

with equality only if $Z = f_j(Z)$.

From (2) it follows that repeated fixing in all indices ad nauseam must lead to a hypergraph $\bar{Z}$ obeying $f_j(\bar{Z}) = \bar{Z}$ for all j , which by (1) has a smaller boundary than Z . ($\bar{Z}$ here is left fixed by all the fixes.)

Now for all but one value of E for each n , $\bar{Z}$ will consist of the E edges with largest θ values proving the theorem where it does so.

How could it fail to do so? Well, it could have some e_1 and not e_2 with $\theta(e_1) > \theta(e_2)$ only if e_1 and e_2 have no vertex in common and lack no vertex in common - for otherwise we contradict that Z is unchanged upon fixing the index in common. Thus, e_1 must be the complement of e_2 . For this to happen $\theta(e_1)$ and $\theta(e_2)$ have to be consecutive among edges: anything in between would be damned if it were in and damned if it were out of Z . Consecutive complements can happen for n even only once, with $\theta(e_1)$ given by $n/2$ followed by a zero then $n/2$ ones and the rest zeros. For n odd it only happens if $\theta(e_1)$ is $(n-1)/2$ followed by $(n+1)/2$ ones and the rest zeros.

For whatever n it is obvious that replacing e_1 by e_2 here reduces the boundary, which proves the theorem in every case.

What can one get out of this theorem?

Define the j-*shadow* of a k-edge to be the j-edges that contain or are contained in it; the j-shadow of a k-hypergraph X is then the union of the j-shadow of its edges. Denote it by $S_j(X)$.

First notice that the theorem tells us how to choose a k-hypergraph having x edges to minimize its (k+1)-shadow. We can choose the one, $H_k(x)$ whose k-edges are those with the largest θ values.

If we could do better with some k-hypergraph Q we could add all j-edges for $j < k$ to Q and obtain a hypergraph of the same size with smaller boundary than the union of these j-edges and $H_k(x)$.

Here are some other consequences.

2. $H_k(x)$ minimizes the j-shadow of an x-edge k-hypergraph for any $j > k$.

3. *The Erdös-Ko-Rado Theorem* For $k \le n/2$ the size of an intersection k-hypergraph Y cannot exceed $\binom{n-1}{k-1}$, the size of the k-hypergraph containing all edges containing one particular vertex.

Proof: No k-edge can both be in Y and be the complement of an edge in the (n-k)-shadow of Y. Thus $Y + S_{n-k}(Y) \le \binom{n}{k}$. If $Y = \binom{n-1}{k-1}$, Y consists of all edges containing the first vertex and this is an equality. The monotonicity of S makes it fail if $Y \ge \binom{n-1}{j-1} + 1$.

4. A slight generalization: If $k_1 + k_2 \le n$ and every member of a k_1-hypergraph Y_1 intersects every member of a k_2-hypergraph Y_2, then $|Y_1| \le \binom{n}{k_1} - S_{n-k_2}(Y_2)$. (Same proof.) This implies a conjecture of Milner that $|Y_1| <$ $\binom{n-1}{j-1}$ or $|Y_2| \le \binom{n-1}{k_2-1}$.

5. There is a straightforward expression for $S_j(Y)$ for a k-hypergraph Y. It is too ugly to repeat, but is straightforward to deduce.

6. If $n/2 \geq j > k$ then $|S_j(Y)| \geq |Y|$ for any k-hypergraph Y .

This follows easily from 5.

7. *Sperner's Theorem.* By 6. given an antichain we can replace its smallest edges by edges one rank (k value) higher and increase its size, if the smaller rank is below $n/2$; and do the complementary thing if the highest rank is above $n/2$. Iterating tells us that largest antichain can have all elements of size $[n/2]$.

(This proof is like using an elephant gun to shoot a flea - there are much easier proofs of Sperner's theorem.)

8. Suppose we have a sequence of integers $\alpha_1 \ldots \alpha_n$. Does there exist an antichain having α_1 k-edges for each k ?

From our results we could prove that there is one if and only if there is one which has as k-edges the α_k with largest value among those not in the k-shadow of smaller edges. By 5. this exactly answers this question albeit in inconvenient form.

9. Given the size of an antichain how small can the ideal of edges contained in its edges be?

Whole classes of such questions can be answered from this theorem. We shall omit all details, and press on.

Incidentally, *the shadow theorem* was proven by J. Kruskal and slightly later by G. Katona. Their proofs were horrendous! It is a kind of Macauley theorem. Nice proofs were supplied by Lindstrom, and generalizations were found by Clements and Lindstrom. Harper, and Wang and Wang proved isoperimetric theorems probably among others, as has Ahlswede and Katona. There exist many variations and generalizations.

Because the purpose of this section is to give you a feeling and not put you to sleep we will now go on to some newer results.

III. RECENT RESULTS

1. Kneser's Conjecture: It is quite an old conjecture that it requires $n - 2k + 2 = r$ intersecting k-hypergraphs on n vertices (with $n \geq 2k-1$) if their union is to contain all k-edges. Recently Lovász proved this, and then Barany found an easy proof. It is easy, as we shall see, but it makes use of two results from other branches of mathematics. In particular it makes use of a theorem of Borsuk that states: if the points of the unit disc in $r-1$ dimensions are coloured in $(r-1)$ colours the diameter of the closure of the points of some colour is 2. It also uses a theorem of Gale, that one can distribute $2k + r - 2$ points on the unit disc in $r-1$ dimensions so that every half containing the origin contains at least k points in its interior.

Now if Kneser's conjecture were wrong, one could cover all k-edges with $r-1$ intersecting hypergraphs. Associate a colour with each hypergraph. We can then associate a colouring of the disc in question in $r-1$ colours by colouring x in any colour of any k-edge lying in the interior of the half plane whose boundary is normal to the origin-to-x vector. But then by Borsuk's theorem the diameter of some colour is 2 which means that that colour contains two disjoint k-edges which violates our definitions. (All of this is easy to visualize if $r = 3$. Then Gale's theorem says that we can put $2k + 1$ points around a unit circle so every half circle contains at least k points in its interior. This is easy - just spread them uniformly. Associate the colour of a k-edge in the half circle centred about x to x. By Borsuk's theorem if the half circle is 2 coloured it contains two points as close to diametrically opposite as one wants in one colour. We can always therefore make the corresponding k-edges disjoint. Things work out exactly the same in more dimensions, though now Gale's theorem requires some proof.)

2. A conjecture of Erdös on linear combinations of n distinct numbers with fixed sums. This is a problem that can be phrased to sound like a hypergraph extremal problem, or not. Here is what it sounds like in hypergraph language.

Suppose we give a distinct weight to each of n vertices, and weigh each edge of the complete hypergraph by the sum of the weight of its vertices. Define $H(x)$ to have as edges all edges that have weight x. How large can $\max |H(x)|$ be?

Many years ago Erdös suggested that the largest value was obtained by having as weights n integers of minimum absolute value.

Recently, R. Stanley proved a result about hypergraphs which comes close to proving this. G.W. Peck proved Erdös' conjecture from Stanley's result. If we describe a hyperedge or set by a sequence of zeros and ones (an incidence sequence), the usual containment ordering has one edge bigger than another if every term in its incidence sequence is larger than the corresponding term in the other's. Now let the derived sequence of a hyperedge have as k-th entry the sum of the first k entries of its incidence sequence.

The derived ordering on hyperedges then has one edge bigger than another if every term in its derived sequence is larger than the corresponding term in the other's.

A derived antichain is a hypergraph no two of whose edges are ordered in the derived ordering. The derived rank of a hyperedge is the sum of the entries of its derived sequence.

Stanley showed that there is a largest derived antichain consisting of all edges having one derived rank (this is a "derived" Sperner theorem) and further that a largest union of k derived antichains consists of all edges having certain k consecutive derived ranks.

If the weights of every vertex are positive definite as well as distinct and are indexed in decreasing order, then $H(x)$ has to be a derived antichain and the maximum size of $H(x)$ is at most the number of hyperedges of some derived rank. If the weight of vertex j is $n+1-j$ one actually attains this maximum. Similar remarks hold for the number of edges having at most k distinct weights - these correspond to the union of k derived antichains. Peck showed that one could carry this result over to the general weight problem to yield Erdös conjecture - but not the k distinct weight analogue, which remains open.

You will be spared the arguments here. Stanley's made use of techniques and results of algebraic geometry. Peck's is left as an assignment for you.

3. Erdös once asked, what is the maximum size of the union of k intersecting hypergraphs on n vertices? He conjectures that it must leave out 2^{n-k} edges out of the 2^n edges that are in the complete hypergraph.

This was proven in 1966 by use of the following lemma: If A and B are ideals than $|A||B| \leq 2^n|A \cap B|$ which is very easy to prove by induction. Then Erdös' conjecture followed, applying this to maximal intersecting hypergraphs, which are complements of ideals. It says that a new intersecting hypergraph must contain at least the same proportion of what was in the others as it does of new edges, which proves the conjecture.

Now it turns out that there are several inequalities

along the same lines that are very useful and important in statistical mechanics. To see a connection, imagine that A and B are random variables that take the value 1 on ideals and zero elsewhere, each edge having equal probability. This inequality then says that the probability of A given B is greater than the probability of A - merely from the fact that A and B are ideals.

A well-known inequality of Kasteleyn-Ginibre, and Fortuin, is the extension of this same statement to the case in which nonnegative weight is associated with each edge.

It takes the form

$$\Big(\sum_{e\in A} \alpha(e)\Big(\sum_{e\in B} \alpha(e)\Big) \le \Big(\sum_{e}\alpha(e)\Big)\Big(\sum_{e\in A\cap B} \alpha(e)\Big)$$

so long as $\alpha(e)\alpha(f) \le \alpha(e\cup f)\alpha(e\cap f)$ and has obvious application to probability. Holley found another generalization. If we have 2 weights α and β on edges and for all edges e and f $\alpha(e)\beta(f) \le \alpha(e\cup f)\beta(e\cap f)$ holds, then the B weight $\Sigma\beta(e)$ of any ideal is larger than its A weight $\Sigma\alpha(e)$. The original proofs of these latter results were quite nontrivial.

Ahlswede and Daykin fairly recently noticed that if one generalized the situation still further, introducing four weights $\alpha,\beta,\gamma,\delta$ with, for all edges e and f ,

$$\alpha(e)\beta(f) \le \gamma(e\cup f)\delta(e\cap f)$$

Then simple induction (as proves the lemma) can be used to obtain that for any two hypergraphs A and B

$$\Big(\sum_{e\in A} \alpha(e)\Big)\Big(\sum_{e\in B} \beta(e)\Big) \le \Big(\sum_{\substack{h=e\cup f\\ \text{for some}\\ e\in A,\, f\in B}} \gamma(h)\Big)\Big(\sum_{\substack{h=e\cap f\\ \text{for some}\\ e\in A,\, f\in B}} \delta(h)\Big)$$

which generalizes all manner of inequalities. To prove it

one simply suppresses one vertex and applies the result inductively to the set of weights that are sums of the given weights over the two edges that project onto each other under that suppression that lie in the appropriate hypergraph. It is a fairly easy matter to verify that the required inequality $\alpha'(e)\beta'(f) \le \gamma'(e'\cup f')\delta'(e'\cap f')$ holds for the summed weights if it held before summing which is all there is to prove.

Daykin has carried this idea one step further obtaining a result that is almost too general to say in ordinary language. We give just one example of its application. Denote the symmetric difference of two edges e, f by $e\Delta f$. Then if for all edges e, f $\alpha(e)\beta(f) \le \gamma(e\Delta f)\delta(e\cap f)$ then for all hypergraphs A, B on the given vertices we have

$$\Big(\sum_{e\in A} \alpha(e)\Big)\Big(\sum_{f\in B} \beta(f)\Big) \le \big(\tfrac{1}{2}(1+\sqrt{2})\big)^n \Big(\sum_{\substack{g = e\Delta f \\ \text{for some} \\ e\in A, f\in B}} \gamma(g)\Big)\Big(\sum_{\substack{h = e\cap f \\ \text{for some} \\ e\in A, f\in B}} \delta(h)\Big)$$

This is the same as before in other words with union replaced by symmetric difference and a factor added.

4. Daykin, D. West and myself have addressed the question: how sharp is that original equality $|A||B| \le 2^n|A\cap B|$? Using the general approach used to obtain the classical theorem of Section 2 we found exactly which hypergraphs of given sizes have the largest intersection; we then showed that the exact bound differed from the one here by at most n. That is, you can actually find A and B so that

$$|A||B|/2^n \ge |A\cap B| + n/4$$

for some ideals A and B. Thus the inequality above is remarkably sharp given its simplicity (imagine for example $n = 100$, $A = 2^{70}$, $B = 2^{90}$; then there are A and B with intersections having size between 2^{60} and $2^{60}+25$). We also find various weighted and multigraph generalizations.

5. Let the k parameter h_k of a hypergraph H be the number of k edges in it. There is a classical inequality satisfied by the parameters of an antichain observed by Yamomoto, Lubell and also Meshalkin, called the LYM inequality, which is

$$\sum_{k=0}^{n} h_k/\binom{n}{k} \le 1$$

(A similar inequality holds for multisets as first observed by Anderson.)

This inequality is quite useful (see for example 6.), implying Sperner's theorem immediately as well as many other things, as described in references [A] and [B].

The inequality is an equality if one has h_k is $\binom{n}{k}$ but is not otherwise. The Kruskal-Katona theorem we discussed in section 2 gives a sharp implication about the h_k in every instance, but it is cumbersome to use.

M. Saks and myself have found a simple improvement, and in fact a class of improvements which become more and more complicated as they approach the full implication of the Kruskal-Katona theorem. The result is as follows.

Let k_o be the smallest value of k for which

$$\sum_{j=0}^{k_o} h_j/\binom{n-1}{j-1} > 1 \quad .$$

Then

$$\sum_{j=0}^{k_o} k_o h_j/j\binom{n}{j} + \sum_{j=k_o+1}^{n} (n-k_o)h_j/(n-j)\binom{n}{j} \le 1 \quad .$$

6. How large can an antichain on n vertices be if it consists of pairs of complementary edges, that is if together with each edge the complement is also an edge of the antichain? The answer, given by Bollobás, is that the largest such antichain has at most $2\binom{n-1}{[n/2]-1}$ edges. Of course, by LYM one need worry only about the case n odd.

A similar question was solved by Clements. How large can an antichain be if it does not contain complementary edges? The answer is $\binom{n}{[(n-1)/2]}$. Here by LYM only the case n even has to be proved. This is done by invoking the Kruskal-Katona theorem and then shifting large edges into smaller ones and then these into $((n/2)-1)$-edges. A generalization to multi-sets, which can be proven the same way, was observed by Gronau.

7. There has been a considerable amount of effort and success recently in the direction of refining the Erdös-Ko-Rado theorem. That theorem says that if you want a maximal intersecting k-hypergraph you can choose all the k-edges containing one vertex. A number of authors have raised questions about what happens if the intersection of all the edges in the hypergraph must be empty, and more recently, if no vertex can appear in more than a proportion α of the edges of the hypergraph.

For $\alpha = 1$ we have the Erdös-Ko-Rado theorem. For somewhat less the best thing to do is was shown by P. Frankl is to choose those k-edges having at least two out of a certain three vertices. For α still smaller, one should choose all k-edges containing an edge in the projective plane on some seven vertices. Recently, Z. Füredi proved that for any α for which one can have a non-empty k-hypergraph, one should find an appropriate projective plane on a subset of vertices and take all k-edges containing an edge of this plane.

There are a number of related problems: Suppose we have an antichain so that every k of its edges have a non-empty intersection. How large can it be? Gronau has precise results for most values of k and n. Other related results have been proved by Erdös, Rado, Bollobás and Daykin.

8. Wang has derived a recursive inequality obeyed by the size of the largest hypergraph in n vertices having all unions of edges larger than s and all intersections smaller than $n-r$. D.L. Wang and P. Wang have generalized the Kruskal-Katona or McCauley theorem to multisets with largest index considered to be next to the smallest ("on a torus").

9. Baumert, McElice, Rodemich and Rumsey have strengthened Sperner's theorem by considering probability distribution. Their result led Shearer and myself to the following open question: Let a chain be a hypergraph whose edges are linearly ordered by inclusion. In how many ways can one partition the edges of the complete hypergraph on n vertices into $\binom{n}{n/2}$ chains so that no two edges appear together in more than one chain?

It is not difficult to show that two ways are possible for $n > 1$ (which proves the Baumert et al. results) and we conjecture that $([n/2]+1)$ is possible.

To find two ways one can form the "left" and "right" de Bruyn partitions ordering the vertices in some order and switch the empty set.

The left de Bruyn partition is as follows. With a given order of vertices, denote an edge by its incidence sequence and then in it replace each zero by a left parenthesis and each one by a right parenthesis. Two edges are in the same chain if the parentheses that can close are the same way in both. The right one is obtained by interchanging left and right in the definition. It is easy to see that only the 0 and n-edges are in the same chain in both.

10. There have been lots of generalizations of results for hypergraphs or set systems, to multisets and to other lattices or natural orders in which the same concepts exist namely in which union intersection, rank, order, can be defined. This is the subject matter of a whole other lecture.

11. I repeat my apologies to those whose results are left out here.

IV. SOME OPEN QUESTIONS

1. Chvatal's Conjecture: We have already mentioned it. Let the principal intersecting hypergraph P_j consist of all edges containing the vertex j. The conjecture is that a principal intersecting hypergraph has maximal intersection with any given ideal, among intersecting hypergraphs.

Chvátal showed this a long time ago for any idea I with the property that substituting the first vertex for any other vertex in an edge of I (not containing the first vertex) will give an edge of I. A long time ago, Magnanti and Kleitman showed that a non-principal intersecting hypergraph containing a 2-edge could not be maximal for any ideal violating the conjecture. Berge gave a related theorem at the 1975 meeting, and Daykin and Hilton have found another proof of Berge's result and some generalizations. D.L. Wang has some partial results in particular, if the ideal has vanishing intersection for any triple of maximum edges.

We make the following observations.

A. It is immediate from the lemma in Section III.3 above that an intersecting hypergraph H of j edges has at least j edges contained in its edges and not in it. (One applies it to the ideal H generated and the complement of any maximal intersecting hypergraph containing H.) By Philip Hall's marriage theorem, this means we can match the edges of any in H with edges of I, so that each edge is disjoint from its image, and the image is contained in an edge of H.

If this is applied to the edges of I and H not containing the first vertex, and the first vertex is added to the images we have mapped H into a principle intersecting hypergraph in Chvátal's case.

B. Having failed to resolve this conjecture we suggest a stronger one - maybe this can be proven wrong:

Consider a vector space in which a basis is corresponded to the edges of the complete hypergraph. A vector V is bigger than another W if one can get from W to V by a sequence of steps in each of which one adds $\lambda(\underset{\sim}{u}_e - \underset{\sim}{u}_f)$ for some e and f where $\underset{\sim}{u}$'s are unit vectors, $\lambda \geq 0$ and $e \supset f$. A hypergraph can be represented by a vector in this space - the sum of its edge unit vectors. Here is a:

Strong Chvátal Conjecture: The vector of any maximal intersecting hypergraph is bigger than some linear combination of principal intersecting hypergraphs with positive coefficients summing to one.

It is easy to derive Chvátal's conjecture from this, and to show that it holds for all intersecting hypergraphs containing a 2-edge; also for those containing a 3-edge if the number of its edges containing at most one of the vertices in that edge is no more than 2^{n-3}. In this case the hypergraph lies above a linear combination of the principal hypergraphs of the three vertices in the edge.

Similar but apparently useless results hold for 4-edges, etc. I encourage you to resolve this for the next meeting.

2. A Co-Boundary Problem: How can we choose a k-hypergraph H on n vertices having α edges to minimize the number

of pairs (e,f) both k-edges one in and one not in H, that are 'neighbours' i.e., have symmetric difference containing two vertices only.

Unlike the Kruskal-Katona result - the answer here is not that one order the k-edges somehow and chooses the largest ones on a list. This fails.

On the other hand we can guess what the answer seems to be - but cannot prove it.

Remarks:

1. The problem does not change if we complement the edges of H, changing k to n-k, or if we complement H as a k-hypergraph.

Conjecture: There is an optimal H for any α so that either H or its complement as a k-hypergraph or the complements of the edges of H or the complements of these as an (n-k)-hypergraph do not contain any edges containing the n-th vertex.

This strange sounding statement actually gives a complete answer, if applied indirectly. For any given α it is easy to see which alternative should hold. Let $k \le n/2$ hold. Then if $\alpha \le \binom{n-1}{k-1}$ every member of H should have the n-th element (so that no member of the hypergraph or complements should). If $1/2\binom{n}{k} \ge \alpha \ge \binom{n-1}{k-1}$ then the complement of H should lack edges with these elements; above $1/2\binom{n}{k}$ we complement these statements. (If you can follow this you deserve better.) Once one knows how to handle the n-th vertex, we can handle the (n-1)-th similarly and so construct H if the conjecture holds.

3. The Unions Problem: We call a hypergraph red if whenever it contains two edges f,g and their union the union is trivial - consisting of f or g. Long ago it was shown that the longest red hypergraph can have no more than

$\binom{n}{n/2}(1+O(n^{-\frac{1}{2}}))$ edges. On the other hand, it seems that one can do no better in constructing such a hypergraph than get $\binom{n}{n/2}(1+O(n^{-1}))$ edges.

The difference may seem too small to worry about, and perhaps it is. We mention it because it looks like improving the bound here seems to require a new idea. We would like to see this resolved not so much because we care about it, but in order to see the new idea, whatever it may be.

In the Erdös tradition, apologizing for inflation, we offer $25 for solution of any of these problems. (But not for the open question in III.9.) The last open questions are hard though, not mathematical. One concerns notation. There is apparently no joy that exceeds that of an author in inventing new terminology for these problems. For antichains one has Sperner sets and clusters, for ideals and their complements there are simplicial complexes, hereditary families, filters up-sets and down-sets. There are set systems and hypergraphs k-sets and k-edges. We have attempted to use a particularly unattractive notation throughout, to illustrate the problem. It would be nice if there were a standard language.

Finally, there is the question of how to improve communication in this area. Though the number of people working in it has probably not exceeded 50 in recent years, results are generally rediscovered more often than they are discovered. Work is published in so many different journals and there are such time delays in publication, that it is very difficult as well as boring to appraise oneself of what others have done and practitioners here including myself rarely seem to make much effort to find out if their discoveries are new. Perhaps conferences such as this one are the best solution.

We thank L. Babai for communicating several of these results.

REFERENCES

Ahlswede, R. and Daykin, D.E. (1978). An inequality for the weights of two families of sets, their unions and intersections. *Z. Wahrschein V. Gebiete* 43, 183-185.

Anderson, I. (1976). Intersection theorems and a lemma of Kleitman. *Discrete Math.* 16, 181-185.

Baumert, D., McEliece, E.R., Rodemich, E.R. and Rumsey, H. A probabilistic version of Sperner's theorem. To appear.

Berge, C. (1976). A theorem related to the Chvátal conjecture. Proceedings of 5th British Combinatorial Conf. Aberdeen (1975), eds. C.St.J.A. Nash-Williams and J. Sheehan. *Utilitas* 35-40.

Bollobás, B. Sperner systems consisting of pairs of complementary subsets. *J. Combinatorial Theory* 15 (1973) 363-366.

Bollobás, B., Daykin, D.E. and Erdös, P. Sets of independent edges of a hypergraph. *Quart. J. Math.* Oxford (2) 27 (1976) 25-32.

Bollobás, B. Disjoint triples in a 3-graph with given maximal degree. *Quart. J. Math.* Oxford (2) 28 (1977) 81-85.

Chvátal, V. Intersecting families of edges in hypergraphs having the hereditary property.

Chvátal, V., Klarner, D.A. and Knuth, D.E. Selected combinatorial research problems.

Clements, G. Complement-free antichains. Preprint.

Clements, G. and Lindstrom, B. (1969). A generalization of a combinatorial theorem of Macaulay. *J. Combinatorial Theory* 7, 230-238.

Daykin, D.E. (1976). Poset functions commuting with the product and yielding Chebychev type inequalities. Colloques Int. C.N.R.S. *Problemes Comb. et Theorie des Graphs*, eds., J.C. Bermond et al., 93-98.

Daykin, D.E. (1977). A lattice is distributive iff $|A||B| \le |AB||AB|$. *Nanta Math.* 10, 58-60.

Daykin, D.E. A hierarchy of inequalities. *Studies in*

Applied Math. (to appear).

Daykin, D.E. Functions on a distributive lattice with a polarity. *J. London Math. Soc.* (to appear).

Daykin, D.E. Inequalities among the subsets of a set. *Nanta Math.* (to appear).

Daykin, D.E. and Hilton, A.J.W. Pairings from down-sets in distributive lattices (to appear).

Daykin, D.E., Kleitman, D.J. and West, D.B. The number of meets between two subsets of a lattice. *J. Combinatorial Theory* (to appear).

Erdös, P. (1945). On a lemma of Littlewood and Offord. *Bull. Amer. Math. Soc.* 51, 898-902.

Erdös, P. and Rado, R. Intersection theorems for systems of sets. *J. London Math. Soc.* 35 (1960) 85-90.

Fortuin, C.M., Kasteleyn, P.W. and Ginibre, J. (1971). Correlation inequalities on some partially ordered sets. *Comm. Math. Phys.* 22, 89-103.

Greene, C. and Kelitman, D.J. (1978). Proof techniques in the theory of finite sets. *M.A.A. Studies in Combinatorics*, ed. G.C. Rota, M.A.A. Monograph.

Gronau, H-D. O.F. On Sperner families in which no k sets have an empty intersection. *J. Combinatorial Theory (A)* to appear.

Gronau, H-D. O.F. On Sperner families in which no 3 sets have an empty intersection. *Acta Cybernetica*, to appear.

Gronau, H-D. O.F. On Sperner families in which no k sets have an empty intersection, II and III, preprints.

Harper, L.H. (1974). The morphology of partially ordered sets. *J. Combinatorial Theory* 17, 44-58.

Holley, R. (1974). Remarks on the FKG inequalities. *Comm. Math. Phys.* 36, 227-231.

Katona, G.O.H. (1974). Extremal problems for hypergraphs. *Combinatorics*, ed. M. Hall and J.H. van Line. Mathematical Centre Tracts 56, Amsterdam.

Kleitman, D.J. (1966). Families of nondisjoint subsets. *J. Combinatorial Theory* 1, 153-155.

Kleitman, D.J. (1968). On a conjecture of Milner on k-graphs with non-disjoint edges. *J. Combinatorial Theory* 5, 153-156.

Kleitman, D.J. (1974). On the extremal property of antichains in partial orders. The LYM property and some of its implications and applications. *Combinatorics*, eds. M. Hall and J.H. van Lint. Math. Centre Tracts 55, Amsterdam, 66-90.

Kleitman, D.J. and Magnanti, T.L. On the number of latent subsets of intersecting collections.

Kleitman, D.J. and Saks, M. Stronger forms of the LYM in equality (in preparation).

Kruskal, J. (1963). The number of simplices in a complex. *Mathematical Optimization Techniques*. University of California Press, Berkeley and Los Angeles, 251-278.

Lubell, D. (1966). A short proof of Sperner's theorem. *J. Combinatorial Theory* 1, 299.

Meshalkin, L.D. (1963). A generalization of Sperner's theorem on the number of subsets of a finite set. *Theor. Probability Appl.* 8, 203-204.

Peck, G.W. Erdös conjecture on sums of distinct numbers (to appear).

Seymour, P.D. and Welsh, D.J.A. (1975). Combinatorial applications of an inequality from statistical mechanics. *Math. Proc. Camb. Phil. Soc.* 77, 485-495.

Shearer, J. and Kleitman, D.J. Probabilities of independent choices being ordered.

Sperner, E. (1928). Ein satz uber untermenge einer endlichen mengs. *Math. Z.* 27, 544-548.

Stanley, R.P. Weyl groups, the hard Lefschetz theorem, and the Sperner property (to appear).

Wang, D.A. On systems of finite sets with constraints on their unions and intersections (to appear).

Wang, D.L. and Wang, P. Discrete isoperimetric problems. *SIAM Journal of Applied Mathematics*.

Wang, D.L. and Wang, P. Extremal configurations in a discrete torus and a generalization of the generalized Macaulay theorems (to appear).

Wang, D.L. and Wang, P. On the bandwidth ordering and isoperimetric problems of graphs. *SIAM Journal* (to appear).

Yamamoto, K. (1954). Logarithmic order of free distributive lattices. *J. Math. Soc. Japan* 6, 343-353.

4 · Connectivity and edge-connectivity in finite graphs

W. Mader

In the last twelve years more and more combinatorialists have taken interest in connectivity problems, and therefore some progress has been made, but there are still more unsolved problems than solved ones. We shall confine ourselves here to finite, undirected graphs and only sometimes we shall mention analogous problems for infinite graphs or for digraphs. Most of the connectivity problems for undirected graphs have a counterpart in the directed case. In general, these "directed" problems are more complicated, and so it may happen that a connectivity problem is completely solved for undirected graphs whereas the corresponding problem for digraphs has not even been attacked. (But there are some connectivity problems for digraphs which have no analogue for undirected graphs, as for instance the question answered in [22]. Also the intermediate result of Nash-Williams [54] (cf. also [45]) should be mentioned.) If not otherwise stated, all graphs are supposed to be undirected and finite without multiple edges or loops.

The article is divided into four parts. In part I we determine the maximum number of openly disjoint or edge-disjoint paths joining vertices of a given set. In part II we try to construct all n-connected (n-edge-connected) graphs and collect some properties of n-connected graphs. In part III we look for some special configurations (circuits through given vertices or edges, subdivisions of complete graphs) in n-connected graphs and in part IV we ask for the maximum number of edges a graph with m vertices may have without containing such configurations as vertices $x \neq y$ joined by n openly disjoint (or edge-disjoint) paths or an n-connected (or n-edge-connected) subgraph.

Before continuing we will fix some notations. The graph $G = (V(G),E(G))$ has *vertex set* $V(G)$ and *edge set* $E(G)$ with $|G| := |V(G)|$ the number of vertices and $||G|| := |E(G)|$ the number of edges of G. The edge between the vertices x and y is denoted by $[x,y]$. For $x \in V(G)$ define $N(x;G) := \{y \in V(G) | [x,y] \in E(G)\}$ and let $\kappa(x;G)$ denote the degree of x in G; in a digraph (= directed graph) D we denote the *indegree* and *outdegree* by $\kappa^-(x;D)$ and $\kappa^+(x;D)$, respectively. For $X \subseteq V(G)$ let $G(X)$ denote the *subgraph* of G induced by X and the set of components of G be denoted by $C(G)$. K_n means the complete graph on n vertices and K_n^- arises from K_n by deleting any edge. For a graph G, the *complement* is denoted by $\bar{G}$ and mG is the union of m (vertex)-disjoint copies of G. The *join* of the graphs G and H (which arises from the disjoint union of G and H by adding all edges $[x,y]$ with $x \in V(G)$ and $y \in V(H)$) is denoted by $G+H$. So the complete bipartite graph $K_{n,m}$ may be written $\bar{K}_n + \bar{K}_m$. - The set of positive integers is denoted by N and we define $N_m := \{n \in N | n \leq m\}$. For $n \in N$, let Z_n be the additive group of integers modulo n. The greatest integer not exceeding the real number r and the smallest integer not less than r are denoted by $[r]$ and $\{r\}$, respectively.

I LOCAL CONNECTIVITY

The fundamental theorem on connectivity has been proven by K. Menger [53] in 1927. A path with endpoints x and y is called an *x,y-path* and the maximal number of openly disjoint resp. edge-disjoint x,y-paths in G is denoted by $\mu(x,y;G)$ resp. $\lambda(x,y;G)$. On the other hand, we consider $T \subseteq V(G) - \{x,y\}$ resp. $T \subseteq E(G)$ with the property that x and y belong to different components of $G - T$. We call such sets of vertices resp. edges x,y-separating and define $t(x,y;G) := \min\{|T| \mid T \subseteq V(G) - \{x,y\}$ is x,y-separating$\}$ and $\tau(x,y;G) := \min\{|T| \mid T \subseteq E(G)$ is x,y-separating$\}$. The first half of the following theorem is Menger's result, while the second half is easily derived from the first one.

<u>Theorem 1</u> [53] For distinct nonadjacent vertices x and y of G,

$$\mu(x,y;G) = t(x,y;G) \quad \text{and} \quad \lambda(x,y;G) = \tau(x,y;G) \quad .$$

Of course, for all vertices $x \neq y$ of G, $\mu(x,y;G) \leq \lambda(x,y;G) \leq \kappa(x;G)$. One might suppose that there is always a system of $\lambda(x,y;G)$ edge-disjoint x,y-paths, which contains a subsystem of $\mu(x,y;G)$ openly disjoint paths. But Beineke and Harary gave a counterexample in [1]. In the same paper they proved also a generalization of Theorem 1, which we state in Theorem 2 below without guaranteeing the truth.* They called a pair (t,s) of nonnegative integers a *connectivity pair* for the vertices $x \neq y$ of G, if there are $T \subseteq V(G) - \{x,y\}$ with $|T| = t$ and $S \subseteq E(G)$ with $|S| = s$ so that x and y belong to different components of $G - (T \cup S)$, but for every $T' \subseteq V(G) - \{x,y\}$ and $S' \subseteq E(G)$ with $|T'| \leq t$, $|S'| \leq s$ and $|T'| + |S'| < t + s$, x and y are contained in the same component of $G - (T' \cup S')$.

<u>Theorem 2</u> [1] If (t,s) is a connectivity pair for vertices $x \neq y$ of G, then there is a system of $t + s$ edge-disjoint x,y-paths $P_1,\ldots,P_{t+s}$ in G, so that also $V(P_i) \cap V(P_j) = \{x,y\}$ for all integers i,j with $1 \leq i < j \leq t$.

---

*The proof given in [1] seems to be incorrect.

It is more natural to ask for the maximum number of edge-disjoint x,y-paths, of which at least t paths are openly disjoint. This question is not at all answered by Theorem 2 (contrary to Harary's hope expressed in Theorem 5.15 in [16]). For (t,s) being a connectivity pair for x,y, there may be more than $t+s$ edge-disjoint x,y-paths, of which more than t are openly disjoint.

Whereas $\min\{\kappa(x;G),\kappa(y;G)\}$ may arbitrarily exceed $\mu(x,y;G)$, it is an interesting fact that every finite graph contains vertices $x \neq y$, for which equality holds. In a weaker form this had been conjectured by Pelikán and Pósa on page 258 in [55] and was proved in [38] and more easily in [42].

Theorem 3 [38] Let G be a finite graph with $E(G) \neq \phi$. Then there are adjacent vertices x and y, for which $\mu(x,y;G) = \min\{\kappa(x;G),\kappa(y;G)\}$ holds.

Obviously, Theorem 3 is not true for multigraphs (multiple edges admitted, but no loops), but multigraphs always contain adjacent vertices x and y with $\lambda(x,y;G) = \min\{\kappa(x;G), \kappa(y;G)\}$. Using Lemma 2 in [45], this can be easily shown by inducing the following statement: Let M be a multigraph and $z \in V(M)$. If $\|M-z\| > 0$, then there are adjacent vertices x and y in $M-z$ with $\lambda(x,y;M) = \min\{\kappa(x;M), \kappa(y;M)\}$.

Let $\mu_n(x,y;G)$ denote the maximum number of openly disjoint x,y-paths in G of length not exceeding n, and for nonadjacent vertices x and y of G let us define $t_n(x,y;G) := \min\{|T| \mid T \subseteq V(G)-\{x,y\}$ such that $G-T$ does not contain an x,y-path of length $\leq n\}$. For nonadjacent vertices x and y, $\mu_n(x,y;G) \leq t_n(x,y;G)$, but simple examples show that $\mu_n(x,y;G) < t_n(x,y;G)$ may occur. This problem was examined by Lovász, Neumann-Lara, and Plummer, and they got the following result.

<u>Theorem 4</u> [27] For all nonadjacent vertices x and y and for every positive integer n, the inequality $t_n(x,y;G) \leq [n/2]\mu_n(x,y;G)$ holds.

This estimate is not the best possible in general. For instance, Lovász, Neumann-Lara and Plummer showed $\mu_4(x,y;G) = t_4(x,y;G)$ in [27]. But for each integer $n \geq 2$ they constructed a graph G_n with nonadjacent vertices x and y so that $t_n(x,y;G_n) \geq [\sqrt{(n/2)}]\mu_n(x,y;G) > 0$. - Let $d(x,y;G)$ denote the distance between the vertices x and y in G. For $n = d(x,y;G) \geq 2$, $\mu_n(x,y;G) = t_n(x,y;G)$ was proved independently in [27] and [7]. This means that the maximum number of openly disjoint geodesic x,y-paths in G is equal to the minimum number of vertices $\neq x,y$, necessary to destroy all geodesic x,y-paths of G.

Though we are dealing with finite graphs above all, I will mention a conjecture of Erdös for infinite graphs. Also for infinite graphs G for any nonadjacent vertices x and y the equality $\mu(x,y;G) = t(x,y;G)$ holds (cf. [12]), where $\mu(x,y;G)$ may be any cardinal number. But the following problem is not solved.

<u>Conjecture</u> (P. Erdös) Let x and y be nonadjacent vertices of the (infinite) graph G. Then there is an x,y-separating set $T \subseteq V(G) - \{x,y\}$ and a system $\mathcal{S}$ of openly disjoint x,y-paths, so that each path of $\mathcal{S}$ contains exactly one vertex of T, and $T \subseteq \bigcup_{P \in \mathcal{S}} V(P)$.

In the countable case this problem has been successfully attacked (but not solved) by Podewski and Steffens in [56].

We now consider an $H \subseteq V(G)$ and we call an x,y-path P an *H-path*, if $x \neq y$ and $V(P) \cap H = \{x,y\}$. Let us denote by $\mu(H;G)$ and $\lambda(H;G)$ the maximum number of openly disjoint and edge-disjoint H-paths in G, respectively. We can suppose H to be independent in G, i.e. $\|G(H)\| = 0$. A $T \subseteq V(G-H) \cup E(G)$ is called *totally H-separating*, if $|C \cap H| \leq 1$ for all $C \in \mathcal{C}(G-T)$. Let $t(H;G) := \min\{|T| \mid T \subseteq V(G-H)$ totally separates H in $G\}$ and $\tau(H;G) := \min\{|S| \mid S \subseteq E(G)$ totally separates H in $G\}$. For $H = \{x,y\}$, $\mu(H;G) = \mu(x,y;G)$ and $t(H;G) = t(x,y;G)$, and correspondingly for λ and τ. It is obvious that $\mu(H;G) \leq t(H;G)$ and $\lambda(H;G) \leq \tau(H;G)$, but the inequalities may be strict. For instance, let us consider a bipartite graph G with bipartition H, $V(G)-H$, and every vertex of $V(G)-H$ may have odd degree in G. Then

$$2\lambda(H;G) = \sum_{x \in V(G)-H} (\kappa(x;G)-1) = \tau(H;G) \quad .$$

Similar examples show that $2\mu(H;G) = t(H;G) < 2t(H;G)$ may occur. It was conjectured by T. Gallai [9] that always $2\mu(H;G) \geq t(H;G)$, and modifying this conjecture, L. Lovász [25] conjectured $2\lambda(H;G) \geq \tau(H;G)$. The first conjecture has been proved by Gallai [9] in the case that for all $x \in H$, $\kappa(x,G) = 1$, and the latter one by Lovász [25] for Eulerian graphs. Both the conjectures are easily derived from the following explicit expressions for $\mu(H;G)$ and $\lambda(H;G)$. For stating this result, we need some further notations. For $X \subseteq V(G)$, we set $E(X;G) := \{[x,y] \in E(G) \mid x \in X \wedge y \in V(G)-X\}$, $\kappa(X;G) := |E(X;G)|$, and $b(X;G) := |X \cap \bigcup_{[x,y] \in E(X;G)} \{x,y\}|$.
We consider sets M consisting of disjoint subsets of $V(G)$ with the property that for all $C \in M$, $|C \cap H| = 1$ and $H \subseteq \bigcup_{C \in M} C$; let $\Pi(H;G)$ denote the collection of all these M.
For $M \in \Pi(H;G)$, $\kappa(M;G) := |\bigcup_{C \in M} E(C;G(\bigcup_{C' \in M} C'))|$.

<u>Theorem 5</u> [44] For every $H \subseteq V(G)$,

$$\lambda(H;G) = \min_{M \in \Pi(H;G)} \left(\kappa(M;G) + \sum_{C \in \mathcal{C}(G - \bigcup_{C' \in M} C')} [\tfrac{1}{2}\kappa(V(C);G)]\right) .$$

We shall now determine $\mu(H;G)$ for independent $H \subseteq V(G)$. For this purpose we consider all (ordered) partitions $(C_o, C_1, \ldots, C_n)$ of $V(G) - H$ and call such a partition *admissible* if $(G - C_o) - \bigcup_{i=1}^{n} E(G(C_i))$ does not contain an H-path. Let the set of all admissible partitions of $V(G) - H$ be denoted by $P(H;G)$.

<u>Theorem 6</u> [47] For every independent $H \subseteq V(G)$,

$$\mu(H;G) = \min_{(C_o, \ldots, C_n) \in P(H;G)} \left(|C_o| + \sum_{i=1}^{n} [\tfrac{1}{2} b(C_i; G - C_o)]\right) .$$

It is possible to derive Theorems 5 and 6 from a more general result. Let us consider functions e and d from $V(G)$ to the nonnegative integers. A system S of edge-disjoint paths in G of length at least 1 is called *admissible in* G_e^d (i.e. (G,e,d)), if for every $x \in V(G)$, there are at most $d(x)$ paths of S passing through x and at most $e(x)$ paths of S ending in x. Let $\lambda(G_e^d)$ denote the maximum number of paths in any admissible path system of G_e^d. In [9] Gallai posed the problem of determining $\lambda(G_e^d)$. (The main result of [9] is a solution of the analogous problem which arises if 'closed paths' are allowed (cf. also [49]).) By applying Theorem 6, in [50] we got a rather difficult expression for $\lambda(G_e^d)$ as a minimum of certain values of the partitions of $V(G)$. This result may be considered as a common generalization of Menger's theorem (Theorem 1) and of Tutte's general factor theorem [61].

II GLOBAL CONNECTIVITY

A multigraph G is called *n-connected* resp. *n-edge-connected*, if $|G| \geq \min\{2, n+1\}$ and for all vertices $x \neq y$, $\mu(x,y;G) \geq n$ resp. $\lambda(x,y;G) \geq n$. The largest n for which G is n-connected resp. n-edge-connected is called the *connectivity number* resp. *edge-connectivity number* of G and is denoted by $\mu(G)$ resp. $\lambda(G)$. A subset $T \subseteq V(G) \cup E(G)$ separates G, if $G - T$ is not connected. The fundamental criterion for n-connectivity was found by Whitney (independently of Menger's result (Theorem 1) from which it is easily deduced).

<u>Theorem 7</u> [64] Let G be a graph with $|G| \geq n+1$. Then G is n-connected if and only if for every separating vertex set $T \subseteq V(G)$, $|T| \geq n$.

For n-edge-connectivity an analogous result holds.

An interesting characteristic property of n-connected graphs has been conjectured by A. Frank and S. Maurer, independently, and was proved independently by L. Lovász and E. Györy.

<u>Theorem 8</u> [26] and [66] Let G be a graph with $|G| \geq n+1$. Then G is n-connected if and only if for any distinct vertices $v_1,\ldots,v_n$ and for any partition of $|G|$ into positive integers $m_1,\ldots,m_n$, there is a partition $V_1,\ldots,V_n$ of $V(G)$ such that for each $i \in N_n$, $v_i \in V_i$, $|V_i| = m_i$, and $G(V_i)$ connected.

This nice result seems to be only a beginning in a 'decomposition theory' for n-connectedness. So far as I know, even the following problem is open: Given any n-connected graph G and any $k \in N_n$, is there always a k-connected subgraph H of G so that $G - E(H)$ is (n-k)-connected? For $k = 2$ this was proved as a by-product on p.190 in [40].

It is a natural question to ask for a successive construction of all n-connected or n-edge-connected graphs. The first result of this kind was obtained by Tutte [62]. Before stating this result, we consider the following operation. Let x be a vertex of G with $\kappa(x;G) \geq 2n-2$. We divide $N(x;G)$ into disjoint subsets N_1 and N_2 with $|N_1| \geq n-1$ and $|N_2| \geq n-1$. We add new vertices x_1 and x_2 to $G-x$ and, furthermore, the edges $\{[x_1,x_2]\} \cup \cup_{i=1}^{2}\{[x_i,y] \mid y \in N_i\}$; so we may obtain G'. We say that G' arises from G by n-vertex-splitting. It is easy to see that every graph which arises from an n-connected graph by n-vertex-splitting or by adding edges is n-connected itself. Conversely, for $n = 3$, Tutte showed that starting from simple types we get all 3-connected graphs in this way.

<u>Theorem 9</u> [62] The class of 3-connected graphs is the class of graphs obtained from the wheels by finite sequences of 3-vertex-splittings and edge-additions.

(A *wheel* is a graph consisting of a circuit C and a further vertex adjacent to all vertices of C.)

P. Slater [57] succeeded in constructing all 4-connected graphs from K_5 in a similar way, but he needs three further operations for this construction. For $n \geq 5$, the problem is unsolved in the case of vertex-connectivity.

The problem is easier and solved for all n in the case of edge-connectivity, if we allow multiple edges. It is convenient to consider even pseudographs, i.e. to admit loops also. We define three operations on a pseudograph G.

O_m : Choose m different edges $e_1,\ldots,e_m$ of G, subdivide e_i by a vertex $x_i \notin V(G)$ and identify $\{x_1,\ldots,x_m\}$ to a vertex $x \notin V(G)$.

O_m^+ : Proceed as in O_m, then choose a vertex $y \in V(G)$ and add a new edge joining x and y.

$\mathcal{O}_m^{(2)}$: Proceed as in $\mathcal{O}_m$, thereby constructing G' . Choose m different edges $e_1',\ldots,e_m'$ of G' , not all incident to x , subdivide e_i' by a vertex $x_i' \notin V(G')$, identify $\{x_1',\ldots,x_m'\}$ to a vertex $x' \notin V(G')$, and add a new edge joining x and x' .

If G' arises from a (2m)-edge-connected pseudograph by $\mathcal{O}_m$, then obviously $\lambda(G') \geq 2m$. The following result was announced by L. Lovász at the Symposium on Graph Theory in Prague (1974).

Theorem 10 (L. Lovász) Starting from K_1 , we obtain all (2m)-edge-connected pseudographs by successive addition of edges and repeated application of $\mathcal{O}_m$.

For constructing all (2m+1)-edge-connected pseudographs we need both the operations $\mathcal{O}_m^+$ and $\mathcal{O}_m^{(2)}$. Of course, both these operations applied to (2m+1)-edge-connected pseudographs yield again (2m+1)-edge-connected pseudographs. The converse is also true.

Theorem 11 [45] Starting from K_1 , we obtain all (2m+1)-edge-connected pseudographs by successive addition of edges and repeated application of $\mathcal{O}_m^+$ and $\mathcal{O}_m^{(2)}$.

The crucial point in the proof of Theorems 10 and 11 is the following reduction method. Let h and k be distinct edges, but not loops of the pseudograph G , incident to $z \in V(G)$; h resp. k may join z and u resp. v . Then we denote by G_z^{hk} the pseudograph which arises from $G-\{h,k\}$ by adding a new edge (perhaps a loop) joining u and v . For all vertices $x \neq y$ of $G-z$, obviously $\lambda(x,y;G_z^{hk}) \leq \lambda(x,y;G)$. The following result had been conjectured (in a weaker form) by Lovász in [24] and proved for Eulerian graphs in [25].

Theorem 12 [45] Let z be a nonseparating vertex of the pseudograph G with $\kappa(z;G) \geq 4$, but without a loop at z. Then there exist different edges h and k incident to z, so that for all vertices $x \neq y$ of $G-z$,

$$\lambda(x,y;G_z^{hk}) = \lambda(x,y;G) \quad .$$

A multigraph G is called *minimally n-edge-connected*, if $\lambda(G) \geq n$, but for all $e \in E(G)$, $\lambda(G-e) < n$. In the proof of Theorems 10 and 11 we also made use of the fact that every minimally n-edge-connected multigraph contains at least two vertices of degree n. This result is easily proved by induction (cf. Lemma 13 in [45]) and follows also from the algorithm of Gomory and Hu [10] (cf. also Ch. IV, §3 in [8]). For minimally n-edge-connected graphs the existence of a vertex of degree n was first explicitly proved by Lick in [21] and generalized by myself in [31].

Theorem 13 [31] Every minimally n-edge-connected graph G contains at least $n+1$ vertices of degree n.

The initiating theorem for these results was given by R. Halin in 1969. He considered *minimally n-connected* graphs, i.e. graphs G with $\mu(G) \geq n$, but for all $e \in E(G)$, $\mu(G-e) < n$.

Theorem 14 [13] Every minimally n-connected graph contains a vertex of degree n.

Notice that Theorem 3 also generalizes this important result.

Let be $V_n(G) := \{x \in V(G) \mid \kappa(x;G) = n\}$ and $|G|_n := |V_n(G)|$. Modifying the methods of Halin, it was possible to get stronger results on $|G|_n$.

Theorem 15 [34] For every minimally n-connected graph G , $|G|_n \geq n+1$.

The following fact implies a good lower bound for $|G|_n$.

Theorem 16 [34] Every circuit in a minimally n-connected graph contains a vertex of degree n .

Hence for every minimally n-connected graph G , the components of $G - V_n(G)$ are trees; let T be such a tree. It is shown in [48] that then for every $x \in V_n(G)$, $|N(x;G) \cap T| \leq n-2$, if $n \geq 3$. This generalizes a result of Halin [14], which says that every circuit of a minimally 3-connected graph contains at least two vertices of degree 3 .

The problem to determine the number of vertices of degree n a minimally n-connected graph must contain is nearly solved.

Theorem 17 [48] For all integers $g > 2n$, $\{\frac{n-1}{2n-1} g + \frac{2n}{2n-1}\} \leq \min\{|G|_n \mid G \text{ minimally n-connected with } |G| = g\} \leq \{\frac{n-1}{2n-1} g + \frac{2n}{2n-1}\} + 1$.

For all $g \equiv 1,3,5,\ldots,2n-1 \pmod{2n-1}$ and $g \equiv 2 \pmod{2n-1}$, the lower bound given in Theorem 17 is the best possible. For $g \equiv 4 \pmod{2n-1}$ and $n \geq 3$, equality holds in the right inequality of Theorem 17. The cases $g \equiv 6,8,\ldots,2n-2 \pmod{2n-1}$ remain undecided.

Whereas, in general, problems for edge-connectivity are easier than the corresponding problems for (vertex-) connectivity, the number of vertices of degree n a minimally n-edge-connected graph must contain, has not been exactly determined up to now. It was shown in [41] that for $n \neq 1,3$, there is a constant $c_n > 0$ so that for all minimally n-edge-connected graphs G, $|G|_n \geq c_n |G|$; the value of c_n was improved in [3]. However this lower bound given in [3] is not believed to be the best possible.

Some of the results on minimally n-connected graphs have been generalized by Y. Hamidoune to directed graphs. For instance, he generalized Theorems 15 and 16.

Theorem 18 [15] For every minimally (strongly) n-connected digraph D, $|\{x \in V(D) | \kappa^+(x;D) = n \vee \kappa^-(x;D) = n\}| \geq n+1$.

Theorem 19 [15] Let H be a (non-empty) subdigraph of the minimally n-connected digraph D, so that for all $x \in V(H)$, $\kappa^+(x;H) \geq 2$ and $\kappa^-(x;H) \geq 2$. Then there is an $x \in V(H)$ so that $\kappa^+(x;D) = n$ or $\kappa^-(x;D) = n$.

These results deal with vertices of indegree or outdegree equal to n. But it is not even known if every minimally n-connected digraph always contains a vertex of, say, out-degree n. I should conjecture more.

Conjecture Every minimally n-connected digraph D contains at least two vertices x with $\kappa^+(x;D) = \kappa^-(x;D) = n$.

For edge-connectivity I could settle this question.

Theorem 20 [39] Every minimally n-edge-connected digraph D contains at least two vertices x with $\kappa^+(x;D) = \kappa^-(x;D) = n$.

One can find examples of minimally n-(edge-) connected digraphs D with arbitrarily large $|D|$ and with $|\{x \in V(D) | \kappa^+(x;D) = \kappa^-(x;D) = n\}| = 2$.

It is possible to generalize Theorem 15 also in another

direction. A subset $H \subseteq V(G)$ is called *n-connected in* G, if for every $x \neq y$ in H, $\mu(x,y;G) \geq n$, and H is called *minimally n-connected in* G, if it is n-connected in G, but for any $e \in E(G)$, it is not n-connected in $G-e$. It was shown in [46] that if H is minimally n-connected in G and $|H| > n$, then $|H \cap V_n(G)| \geq n+1$. But there are also some problems for 'relative n-connectivity' which have no counterpart for n-connectivity. Let H be any finite set and let us consider the class $\mathcal{G}_n(H)$ of all connected graphs G such that H is minimally n-connected in G. If $G \in \mathcal{G}_n(H)$ then every subdivision of G also belongs to $\mathcal{G}_n(H)$. But the following theorem implies that $\mathcal{G}_n(H)$ contains only a finite number of non-homeomorphic graphs.

<u>Theorem 21</u> [46] If H is minimally n-connected in G, then

$$|\{x \in V(G) \mid \kappa(x;G) \geq 3\}| \leq 2^{n+3} n^{2n+2} |H|^2$$

The following concept is closely related to minimal n-connectivity. A graph G is called *critically n-connected*, if it is n-connected, but for each $x \in V(G)$, $\mu(G-x) < n$. Let m,k be positive integers, where $m \geq 4$. Take m disjoint copies K_k^i of K_k $(i \in Z_m)$ and form the graph

$$G_m(k) := \bigcup_{i \in Z_m} (K_k^i + K_k^{i+1})$$

Then $G_m(k)$ is critically (2k)-connected and regular of degree $3k-1$, and $K_1 + G_m(k)$ is a critically (2k+1)-connected graph of minimum degree $3k$. These examples show that the bound given in the next theorem, which was found by Chartrand, Kaugars and Lick in [4] and more easily proved in [30], is the best possible.

Theorem 22 [4] Every critically n-connected (finite) graph G contains a vertex x with $\kappa(x;G) \le [\frac{3n}{2}] - 1$.

The following concept suggests itself, and was introduced by Maurer and Slater in [51]. A graph G is called *k-critically n-connected* (or, briefly, an *(n,k)-graph*), if for every $V' \subseteq V(G)$ with $|V'| \le k$, we have $\mu(G-V') = n - |V'|$. For $k = 1$, we regain the concept of critical n-connectivity. The complete graph K_{n+1} is n-critically n-connected, and it is not difficult to see that it is the only (n,n)-graph (Proposition 2.1 in [51]). Let F_n be a 1-factor of K_{2n} and define $S_n := K_{2n} - E(F_n)$. Then S_{n+1} is a (2n,n)-graph, but it is not (n+1)-critical. This led Slater to the following interesting

Conjecture (P. Slater [51]) With the only exception of K_{n+1}, there is no (n,k)-graph with $k > [\frac{n}{2}]$.

In [51] Maurer and Slater proved this conjecture for all $n \le 6$, and I succeeded later in proving it for all $n \le 10$.

The graph $H_m := (Z_m, \{[x,x+\kappa] | x \in Z_m \wedge (\kappa = 1 \vee \kappa = 2)\})$ is for all $m \ge 5$ a (4,2)-graph. (It is possible to show that there are exactly two further (4,2)-graphs.) From this it follows easily that for each $m > n \ge 4$, there is an (n,2)-graph containing m vertices. But considering (n,3)-graphs this situation changes.

Theorem 23 [43] Every (n,3)-graph contains at most $6n^2$ vertices.

For $m = 3$, I could prove the following conjecture, which would obviously imply Slater's conjecture.

Conjecture For $m \geq 3$, the only $(2m,m)$-graphs are K_{2m+1} and S_{m+1}.

Also the following conjecture is stronger than Slater's one, and I could prove it for all $n \leq 10$.

Conjecture Every $(n,[\frac{n}{4}]+1)$-graph contains a vertex of degree n.

(Cf. also Conjecture 2 in [6].)

Let us deduce Slater's conjecture from ours. If there were a non-complete $(2n+1,n+1)$-graph G, then the graph $G[K_2]$ arising from G by replacing each vertex by two adjacent vertices would be a $(4n+2,n+1)$-graph without vertices of degree $4n+2$. But if there is no $(2n+1,n+1)$-graph different from K_{2n+2}, then also there is no $(2n+2,n+2)$-graph different from K_{2n+3}.

For the corresponding concepts of k-minimal n-connectivity resp. n-edge-connectivity the problems are much simpler. By applying Theorem 15 the 2-minimally n-connected graphs are easily determined: they exist only for $n = 2$ and are the circuits, as found independently in [43] and [52]. Admitting multiple edges in the case of edge-connectivity, one gets all p-circuits C_n^p, which arise from the circuits C_n by replacing each edge by p parallel edges.

Theorem 24 [52] There are k-minimally n-edge-connected multigraphs containing at least 3 vertices only for $k = 2$ and n even, and these are the generalized circuits $C_m^{n/2}$.

There has also been considered a concept which lies between 2-minimal n-connectivity and 2-critical n-connectivity. In [36] all graphs have been determined which have the property that the connectivity number decreases by 2, if we delete any vertex x and any edge not adjacent to x .

III CERTAIN CONFIGURATIONS IN n-CONNECTED GRAPHS

Let G be an n-connected graph with $n \geq 2$. It is a classical result by Dirac [5] that there is a circuit through any given n vertices. Let H be any set of $n+1$ vertices of G . Then there is not necessarily a circuit containing H . Necessary and sufficient conditions for H being contained in a circuit have been found by Mesner and Watkins.

<u>Theorem 25</u> [63] Let G be an n-connected graph with $n \geq 3$ and be $H \subseteq V(G)$ with $|H| = n+1$. Then there is a circuit in G containing H if and only if there is no totally H-separating set $T \subseteq V(G) - H$ with $|T| = n$.

For $n = 2$, Mesner and Watkins have also given necessary and sufficient conditions in [63], but these are more involved.

If there are adjacent vertices x and y in H , then Theorem 25 shows the existence of a circuit containing H . It is easily seen also that there is even a circuit C with $H \subseteq V(C)$ and $[x,y] \in E(C)$. It has been conjectured by Lovász and Woodall, independently, that in an n-connected graph there is a circuit through any given n independent (i.e. vertex-disjoint) edges.

<u>Conjecture</u> (L. Lovász [23], D.R. Woodall [65]) Let G be an n-connected graph with $n \geq 2$. Given any n independent edges $e_1,\ldots,e_n$, there is a circuit containing $e_1,\ldots,e_n$, unless n odd and $G - \{e_1,\ldots,e_n\}$ disconnected.

For $n \leq 3$, the conjecture is easily proved. For arbitrary n , the only progress I know was made by Woodall in

[65], improved later by Thomassen in [60].

Let us now consider disjoint vertex sets $X = \{x_1,\dots,x_n\}$ and $Y = \{y_1,\dots,y_n\}$ in an n-connected graph G. From Theorem 1 one gets immediately the existence of n disjoint x,y-paths in G with $x \in X$ and $y \in Y$. But it is not certain that there are disjoint x_i,y_i-paths in G for $i = 1,\dots,n$. Let us call a graph G *n-linked*, if $|G| \geq 2n$ and for every sequence $x_1,\dots,x_{2n}$ of $2n$ distinct vertices of G, there are disjoint x_i,x_{n+i}-paths for $i = 1,\dots,n$ in G. It is straightforward that for every n-linked graph G, $\mu(G) \geq 2n-1$. But a (2n-1)-connected graph is not necessarily n-linked. For instance, take any 5-connected plane graph G containing a circuit C_4 which is the boundary of a face (such graphs exist). If x_1,x_2,x_3,x_4 are the vertices along C_4, then there are no disjoint x_i,x_{i+2}-paths for $i = 1,2$. Hence G is not 2-linked. But every non-planar 4-connected graph is 2-linked, as shown by Jung in [17]. It is a natural question, if for each n there is a (smallest) integer $f(n)$ such that $\mu(G) \geq f(n)$ implies G to be n-linked. The above mentioned result shows $f(2) = 6$. (For infinite graphs such a function does not exist, because there are infinite planar graphs of arbitrarily great (finite) connectivity and without triangles, as displayed in Fig.8 in [29].) The existence of such a function f for finite graphs has been proved by Larman and Mani [18] and Jung [17], independently.

Theorem 26 ([18] and [17]) If $\mu(G) \geq 2^{3n}$, then G is n-linked.

Proving this theorem, they got the following result (explicitly in [17] and implicitly in [18], too):

(Z) Let G be a (2m)-connected graph containing a subdivision of K_{3m}. Given any multigraph M without isolated vertices and with $||M|| = m$ and $V(M) \subseteq V(G)$, then by suitably subdividing the edges of M one can get a subgraph of G .

For deriving Theorem 26 from (Z) we need a criterion for the existence of subdivisions of K_n in a graph. Such a criterion was first given in [28] and improved in [35].

Theorem 27 [35] If $||G|| \geq 2^n|G|$, then G contains a subdivision of K_{n+1} .

The value 2^n in this theorem (and therefore also the bound in Theorem 26*) is not believed to be the best possible. But it is easily seen that it is at least of order n^2 . For positive integers m,k , take $m+1$ disjoint sets $M_o,M_1,\ldots,M_m$ with $|M_o| = |M_1| = \ldots = |M_m| = k$ and consider the graph

$$G_k^m := (\bigcup_{i=0}^{m} M_i, \bigcup_{i=1}^{m} \{[x,y] | x \in M_{i-1} \wedge y \in M_i\}) .$$

Let $n \geq 2$ be any positive integer and let $\dot{K}$ be any subdivision of K_s contained in $G^m_{\binom{n}{2}}$. The set of principal vertices (branch-vertices) of $\dot{K}_s$ may be denoted by B . Then we have $|B \cap M_i| < 2n$ for all $i = 0,1,\ldots,m$, and for each $i \in N_{m-1}$,

$$|B \cap \bigcup_{j=0}^{i-1} M_j| < n \quad \text{or} \quad |B \cap \bigcup_{j=i+1}^{m} M_j| < n .$$

These facts imply $s = |B| < 6n$. Hence for every $m \in N$ there is a graph G (of order $(m+1)\binom{n}{2}$) with $\frac{||G||}{|G|} = \frac{m}{m+1}\binom{n}{2}$, which does not contain a subdivision of K_{6n} . If therefore

---

*The function f considered in the paragraph before Theorem 26 was conjectured by M. Rosenfeld to be $f(n) = \binom{2n}{2}$. This (unlikely) conjecture was reported by Grünbaum in [11].

c_n and a_n are constants such that every graph G with $||G|| \geq c_n|G| + a_n$ contains a subdivision of K_n, then $c_n \geq \binom{[n/6]}{2}$.

If we take $M = K_n$ in (Z) and combine (Z) and Theorem 27, we get

Theorem 28 ([18] and [17]) If $\mu(G) \geq 2^{3\binom{n}{2}}$, then for every distinct vertices $b_1,\ldots,b_n$ in G, there is a subdivision of K_n in G having $b_1,\ldots,b_n$ as principal vertices.

Trying to give a direct proof of Theorem 28 without using Theorem 27, Lovász posed the following conjecture.

Conjecture (L. Lovász) For each $n \in N$ there is a (smallest) $g(n) \in N$ with the following property: Given any $g(n)$-connected graph G and any distinct vertices x and y of G, there is an x,y-path P in G such that $\mu(G-V(P)) \geq n$.

It is easily seen by induction that $g(1) = 3$, and Lovász proved his conjecture for $n = 2$, too (oral communication).

IV EXTREMAL CONNECTIVITY PROBLEMS

We shall consider the question how many edges (dependent on the number of vertices) a graph may have without containing such configurations as two vertices joined by n openly disjoint (or edge-disjoint) paths or an n-connected (or n-edge-connected) subgraph. These problems are settled completely but for edge-connectivity.

Let G be a graph, all vertices of which have degree $n-1$ except just for one vertex, which is adjacent to all other vertices. Then we have $||G|| = \frac{n}{2}(|G|-1)$ and obviously there are no vertices x and y with $\lambda(x,y;G) \geq n$. B. Bollobás conjectured that every graph G with $||G|| > \frac{n}{2}(|G|-1)$

contains vertices x and y with $\lambda(x,y;G) \geq n$. For $n=4$ this is a consequence of the results in [2], for $n = 5$ and $n = 6$ it was proved in [19] and [20], and the general case was settled in [37] by proving the following generalization.

Theorem 29 [37] Every graph G with

$$||G|| > \frac{n}{2}(|G|-1) - \frac{1}{2} \sum_{x \in V(G) \wedge \kappa(x;G)<n} (n-1-\kappa(x;G))$$

and $|G| \geq n$ contains vertices $x \neq y$ with $\lambda(x,y;G) \geq n$.

Whereas by [2] every 2-connected graph G with $||G|| = 2(|G|-1)$ and $\max_{x\neq y} \lambda(x,y;G) < 4$ is of the type given above (namely a wheel for $n = 4$, if 2-connected), I have not succeeded in characterizing the 'extremal graphs' for $n \geq 5$. But in an extremal graph G with $||G|| = \frac{n}{2}(|G|-1)$ every least separating edge set T has a simple structure: It is $|T| = n-1$ and there is an edge $[t,s] \in T$ with $\{t,s\} \cap \{t',s'\} \neq \phi$ for all $[t',s'] \in T$. By splitting G 'along this edge $[t,s]$' we get two extremal graphs again.

The corresponding problem for openly disjoint paths is solved only for $n \leq 5$. For $n \leq 4$ the solution is the same as for edge-disjoint paths [2]. For $n \geq 5$ let us consider a graph G with $\bar{\mu}(G) := \max_{x\neq y} \mu(x,y;G) < n$, which contains different edges $[a,b]$ and $[a',b']$ with $\mu(a,b;G) \leq \frac{n}{2}$ and $\mu(a',b';G) \leq \frac{n+1}{2}$. We take m disjoint copies $G_1,\ldots,G_m$ of G, where the vertices a_i,b_i,a_i',b_i' of G_i may correspond to the vertices a,b,a',b', respectively. The graph H arises from $G_1,\ldots,G_m$ by identifying a_i with a_{i+1}' and b_i with b_{i+1}' for each $i \in N_{m-1}$. Then $\bar{\mu}(H) < n$, because $\mu(a_i,b_i;H) \leq [\frac{n}{2}] + [\frac{n+1}{2}] - 1 = n-1$ for all $i \in N_{m-1}$. We have $|H| = m(|G|-2) + 2$ and $||H|| = m(||G||-1) + 1$, therefore $||H|| = \frac{||G||-1}{|G|-2}(|H|-2) + 1$. If $G = K_1 + G'$, where G' is an $(n-2)$-regular graph, then

$\|G\| = \frac{n}{2}(|G|-1)$ and therefore $\|H\| = \frac{\frac{n}{2}(|G|-1)-1}{|G|-2}(|H|-2)+1$. For $n = 5$ let G' arise from two disjoint copies K_4^1 and and K_4^2 of K_4 by subdividing exactly one edge $[x_i,y_i]$ of K_4^i by one vertex z_i for $i = 1,2$ and by adding the edge $[z_1,z_2]$. For $G := K_1 + G'$ then $\mu(z_1,z_2;G) = 2$, $\mu(z_1,x_1;G) = 3$, and $\bar{\mu}(G) < 5$. By the above construction (taking $[a,b] = [z_1,z_2]$ and $[a',b'] = [z_1,x_1]$) we get a graph H of arbitrarily large order $|H|$ with $\bar{\mu}(H) < 5$ and $\|H\| = \frac{8}{3}(|H|-2)+1$. This yields the sharpness of the inequality in the following theorem.

Theorem 30 [58] Every graph G with $\|G\| \geq [\frac{8}{3}|G|] - 3$ and $|G| \geq 13$ contains vertices $x \neq y$ with $\mu(x,y;G) \geq 5$.

For $n \geq 6$ we take two disjoint copies K_{n-1}^1 and K_{n-1}^2 of K_{n-1} and choose different vertices x_i,y_i of K_{n-1}^i for $i = 1,2$. The graph G' may arise from $K_{n-1}^1 \cup K_{n-1}^2$ by deleting the edges $[x_1,y_1]$ and $[x_2,y_2]$ and by adding the edges $[x_1,x_2]$ and $[y_1,y_2]$. For $G := K_1 + G'$ then $\mu(x_1,x_2;G) = \mu(y_1,y_2;G) = 3 \leq \frac{n}{2}$ and $\bar{\mu}(G) < n$. For the graph H arising from G as above, $\|H\| = \frac{n(n-1)-1}{2n-3}(|H|-2)+1$ and $\bar{\mu}(H) < n$. (A related construction is found in [58], Cor.2.) I should conjecture that $\frac{n(n-1)-1}{2n-3}$ is the right value for an analogue to Theorem 30 for $n = 6$. But that is not true of all $n \geq 6$, as for instance a construction given by Bollobás [2a] shows (cf. ch.1, §5 in [2b], too).

Theorem 31 [2a] There is a constant $c > 0$ such that for each $n \geq 5$ there are graphs G with arbitrarily large $|G|$ and with $\max_{x \neq y} \mu(x,y;G) < n$, for which $\|G\| > (\frac{n}{2} + c\sqrt{n})|G|$.

Whereas for the maximum number of edges a graph G with

$\max_{x \neq y} \mu(x,y;G) < n$ may have, there is not even a conjecture, the following very similar question has a nice answer. If we ask for the maximum number of edges in a graph G with

$$\max_{[x,y]\in E(G)} \mu(x,y;G) < n \quad \text{resp.} \quad \max_{[x,y]\in E(G)} \lambda(x,y;G) < n \ ,$$

both the problems (for openly disjoint paths and for edge-disjoint paths) coincide and are solved rather completely. From Theorem 3 we easily deduce

<u>Theorem 32</u> [32] Every graph G with $\|G\| > n|G| - \binom{n+1}{2}$ and $|G| \geq n$ contains adjacent vertices x and y with $\mu(x,y;G) > n$.

This result is not sharp for graphs G with $|G| > n+1 \geq 3$. But we are able to show that every graph G with sufficiently large number of vertices and $\|G\| > n|G| - n^2 = \|K_{n,|G|-n}\|$ contains adjacent vertices x and y with $\mu(x,y;G) > n$ and that the extremal graphs of this problem are the complete bipartite graphs $K_{n,m}$.

<u>Theorem 33</u> [32] For each $n \geq 2$ there is an integer $m(n) \leq \binom{n}{2} + 2n+1$ with the property that every graph $G \neq K_{n,|G|-n}$ with $\|G\| \geq n|G|-n^2$ and $|G| \geq m(n)$ contains adjacent vertices x and y with $\mu(x,y;G) > n$.

I should suppose that in Theorem 33 we can take $m(n) = 3n$, which would be the best possible and which I succeeded in proving for edge-disjoint paths.

Let us now consider the question of how many edges a graph without an n-connected (n-edge-connected) subgraph can contain. It is easy to give an answer in the case of edge-connectivity.

Theorem 34 [31] Every graph G with $\|G\| > (n-1)|G| - \binom{n}{2}$ and $|G| \geq n$ contains an n-edge-connected subgraph.

This result is the best possible for all n and every number of vertices. For (vertex-) connectivity we can give a complete solution but for small n and inequalities in the general case. In [33] for each $n \in N$ the existence of a number $g(n)$ was established with the following property: There are infinitely many nonisomorphic graphs G with $\|G\| = g(n)(|G|-(n-1))$ containing no n-connected subgraph, but every graph G with a sufficiently large number of vertices and $\|G\| > g(n)(|G|-(n-1))$ contains an n-connected subgraph. For the graphs $G_m := (mK_{n-1}) + \bar{K}_{n-1}$, $\|G_m\| = (\frac{3}{2}n-2)(|G_m|-(n-1))$, and G_m contains no n-connected subgraph. This gives the lower bound in the following theorem.

Theorem 35 [33] $\frac{3}{2}n-2 \leq g(n) < (n-1)(1+\frac{1}{2}\sqrt{2})$ for all $n \geq 2$.

The upper bound was somewhat improved by Sweetman in [59]. I do believe that in the lower estimate we even have equality.

Conjecture Every graph G with $\|G\| > (\frac{3}{2}n-2)(|G|-(n-1))$ and with sufficiently large $|G|$ contains an n-connected subgraph $(n \geq 2)$.

For all $n \leq 7$, I have proved this conjecture in [33]. For a small vertex number there are graphs G with $\|G\| > (\frac{3}{2}n-2)(|G|-(n-1))$ containing no n-connected subgraph as, for instance, the graphs $(m\bar{K}_{n-1}) + (N_{n-1}, \{[i,n-i] \mid i = 1,\ldots,[\frac{n-1}{2}]\})$ for $m < [\frac{n-1}{2}]$.

REFERENCES

[1] Beineke, L.W. and Harary, F. (1967). The connectivity function of a graph. *Mathematika* 14, 197-202.

[2] Bollobás, B. (1966). On graphs with at most three independent paths connecting any two vertices, *Studia Sci. math. Hungar.* 1, 137-140.

[2a] Bollobás, B. (1978). Cycles and semi-topological configurations, pp.66-74 in *Theory and Applications of Graphs*, Proceedings, Michigan 1976. Springer-Verlag, Berlin.

[2b] Bollobás, B. (1978). *Extremal Graph Theory*. Academic Press, London.

[3] Bollobás, B., Goldsmith, D.L. and Woodall, D.R. Indestructive deletions of edges from graphs. Submitted to *J. of Combinatorial Theory*.

[4] Chartrand, G., Kaugars, A. and Lick, D.R. (1972). Critically n-connected graphs. *Proc. Amer. Math. Soc.* 32, 63-68.

[5] Dirac, G.A. (1960). In abstrakten Graphen vorhandene vollständige 4-Graphen und ihre Unterteilungen. *Math. Nachr.* 22, 61-85.

[6] Entringer, R.C. and Slater, P.J. (1977). A note on k-critically n-connected graphs. *Proc. Amer. Math. Soc.* 66, 372-375.

[7] Entringer, R.C., Jackson, D.E. and Slater, P.J. Geodetic connectivity of graphs. To appear in *IEEE Trans. Circuits and Systems*.

[8] Ford, L.R. and Fulkerson, D.R. (1962). *Flows in Networks*. Princeton University Press, Princeton, New Jersey.

[9] Gallai, T. (1961). Maximum-Minimum-Sätze und verallgemeinerte Faktoren von Graphen. *Acta Math. Acad. Sci. Hungar.* 12, 131-173.

[10] Gomory, R.E. and Hu, T.C. (1961). Multi-terminal network flows. *J. SIAM* 9, 551-570.

[11] Grünbaum, B. (1970). Problem 3 on p.492 in 'Combinatorial Structures and their Applications', *Proceedings of the Calgary International Conference on Combinatorial Structures.* Gordon and Breach, Science Publishers, New York.

[12] Halin, R. (1968). Zum Mengerschen Graphensatz, pp.41-48 in *Beiträge zur Graphentheorie.* Teubner Verlag, Leipzig.

[13] Halin, R. (1969). A theorem on n-connected graphs. *J. Combinatorial Theory* 7, 150-154.

[14] Halin, R. (1969). Untersuchungen über minimale n-fach zusammenhängende Graphen. *Math. Ann.* 182, 175-188.

[15] Hamidoune, Y.O. Quelques problèmes de connexité dans les graphes orientés. To appear in *J. of Combinatorial Theory B.*

[16] Harary, F. (1969). *Graph Theory.* Addison-Wesley Publishing Company, Reading, Massachusetts.

[17] Jung, H.A. (1970). Eine Verallgemeinerung des n-fachen Zusammenhangs für Graphen. *Math. Ann.* 187, 95-103.

[18] Larman, D.G. and Mani, P. (1970). On the existence of certain configurations within graphs and the 1-skeletons of polytopes. *Proc. London Math. Soc.* 20, 144-160.

[19] Leonard, J.L. (1972). On graphs with at most four line-disjoint paths connecting any two vertices. *J. Combinatorial Theory B.* 13, 242-250.

[20] Leonard, J.L. (1973). Graphs with 6-ways. *Can. J. Math.* 25, 687-692.

[21] Lick, D.R. (1972). Minimally n-line connected graphs. *J. Reine Angew. Math.* 252, 178-182.

[22] Lovász, L. (1973). Connectivity in digraphs. *J. Combinatorial Theory* (B) 15, 174-177.

[23] Lovász, L. (1974). Problem 5. *Period. Math. Hungar.* 4, 82.

[24] Lovász, L. (1976). p.684 in *Proceedings of the Fifth British Combinatorial Conference 1975*. Utilitas Mathematica Publishing, Winnipeg.

[25] Lovász, L. (1976). On some connectivity properties of eulerian graphs. *Acta Math. Acad. Sci. Hungar.* 28, 129-138.

[26] Lovász, L. A homology theory for spanning trees of a graph (preprint).

[27] Lovász, L., Neumann-Lara, V. and Plummer, M.D. (1978). Mengerian theorems for paths of bounded length. *Period. Math. Hungar.* 9, 269-276.

[28] Mader, W. (1967). Homomorphieeigenschaften und mittlere Kantendichte von Graphen. *Math. Ann.* 174, 265-268.

[29] Mader, W. (1968). Homomorphiesätze für Graphen. *Math. Ann.* 178, 154-168.

[30] Mader, W. (1971). Eine Eigenschaft der Atome endlicher Graphen. *Arch. Math.* 22, 333-336.

[31] Mader, W. (1971). Minimale n-fach kantenzusammenhängende Graphen. *Math. Ann.* 191, 21-28.

[32] Mader, W. (1971). Existenz gewisser Konfigurationen in n-gesättigten Graphen und in Graphen genügend großer Kantendichte. *Math. Ann.* 194, 295-312.

[33] Mader, W. (1972). Existenz n-fach zusammenhängender Teilgraphen in Graphen genügend großer Kantendichte. *Abh. Math. Sem. Universität Hamburg* 37, 86-97.

[34] Mader, W. (1972). Ecken vom Grad n in minimalen n-fach zusammenhängenden Graphen. *Arch. Math.* 23, 219-224.

[35] Mader, W. (1972). Hinreichende Bedingungen für die Existenz von Teilgraphen, die zu einem vollständigen Graphen homöomorph sind. *Math. Nachr.* 53, 145-150.

[36] Mader, W. (1973). 1-Faktoren von Graphen. *Math. Ann.* 201, 269-282.

[37] Mader, W. (1973). Ein Extremalproblem des Zusammenhangs von Graphen. *Math. Zeitschrift* 131, 223-231.

[38] Mader, W. (1973). Grad und lokaler Zusammenhang in endlichen Graphen. *Math. Ann.* 205, 9-11.

[39] Mader, W. (1974). Ecken vom Innen- und Außengrad n in minimal n-fach kantenzusammenhängenden Digraphen. *Arch. Math.* 25, 107-112.

[40] Mader, W. (1974). Kreuzungsfreie a,b-Wege in endlichen Graphen. *Abh. Math. Sem. Universität Hamburg* 42, 187-204.

[41] Mader, W. (1974). Kantendisjunkte Wege in Graphen. *Monatshefte für Mathematik* 78, 395-404.

[42] Mader, W. (1975). Ecken mit starken Zusammenhangseigenschaften in endlichen Graphen. *Math. Ann.* 216, 123-126.

[43] Mader, W. (1977). Endlichkeitssätze für k-kritische Graphen. *Math. Ann.* 229, 143-153.

[44] Mader, W. (1978). Über die Maximalzahl kantendisjunkter A-Wege. *Arch. Math.* 30, 325-336.

[45] Mader, W. (1978). A reduction method for edge-connectivity in graphs. *Annals of Discrete Mathematics* 3, 145-164.

[46] Mader, W. (1978). Über n-fach zusammenhängende Eckenmengen in Graphen. *J. Combinatorial Theory B.* 25, 74-93.

[47] Mader, W. (1978). Über die Maximalzahl kreuzungsfreier H-Wege. *Arch. Math.* 31, 387-402.

[48] Mader, W. (1979). Zur Struktur minimal n-fach zusammenhängender Graphen. *Abh. Math. Sem. Universität Hamburg* 49, 49-69.

[49] Mader, W. Über ein graphentheoretisches Ergebnis von T. Gallai. To appear in *Acta Math. Acad. Sci. Hungar.*

[50] Mader, W. Über ein graphentheoretisches Problem von T. Gallai. Submitted to *Archiv der Mathematik*.

[51] Maurer, St. B. and Slater, P.J. (1977). On k-critical, n-connected graphs. *Discrete Mathematics* 20, 255-262.

[52] Maurer, St. B. and Slater, P.J. (1978). On k-minimally n-edge-connected graphs. *Discrete Mathematics* 24, 185-195.

[53] Menger, K. (1927). Zur allgemeinen Kurventheorie. *Fundamenta Math.* 10, 96-115.

[54] Nash-Williams, C.St.J.A. (1960). On orientations, connectivity and odd-vertex-pairings in finite graphs. *Can. J. Math.* 12, 555-567.

[55] Pelikán, J. (1968). Valency conditions for the existence of certain subgraphs, pp.251-258 in *Theory of Graphs* (Proceedings of the Colloquium held at Tihany). Akadémiai Kiadó, Budapest.

[56] Podewski, K.-P. and Steffens, K. (1977). Über Translationen und den Satz von Menger in unendlichen Graphen. *Acta Math. Acad. Hungar.* 30, 69-84.

[57] Slater, P.J. (1974). A classification of 4-connected graphs. *J. Combinatorial Theory B.* 17, 281-298.

[58] Sørensen, B.A. and Thomassen, C. (1974). On k-rails in graphs. *J. Combinatorial Theory B.* 17, 143-159.

[59] Sweetman, D.C. The existence of an n-connected subgraph in a graph with many edges (unpublished manuscript).

[60] Thomassen, C. (1977). Note on circuits containing specified edges. *J. Combinatorial Theory B.* 22, 279-280.

[61] Tutte, W.T. (1952). The factors of graphs. *Can. J. Math.* 4, 314-328.

[62] Tutte, W.T. (1961). A theory of 3-connected graphs. *Indag. Math.* 23, 441-455.

[63] Watkins, M.E. and Mesner, D.M. (1967). Cycles and connectivity in graphs. *Can. J. Math.* 19, 1319-1328.

[64] Whitney, H. (1932). Congruent graphs and the connectivity of graphs. *Amer. J. Math.* 54, 150-168.

[65] Woodall, D.R. (1977). Circuits containing specified edges. *J. Combinatorial Theory B*. 22, 274-278.

[66] Györi, E. (1978). On division of graphs to connected subgraphs, pp.485-494 in *Combinatorics*, North-Holland Publishing Company, Amsterdam-Oxford-New York.

5 · Partition theory and its application

Jaroslav Nešetřil and Vojtech Rödl

Contents:

§1 Introduction

Partition theory was motivated by the following three theorems.

I. The Finite Ramsey Theorem (FRT) [9]. Given positive integers k,m,p, there exists a positive integer n such that for every set X with at least n elements the following statement is true. For every partition of the set $\binom{X}{p}$ into k classes there exists a m-element subset $Y \subseteq X$ all of whose p-element subsets belong to one of the classes of the given partition. Here $\binom{X}{p}$ denotes the set of all p-element subsets of the set X (sometimes we denote this set by $[X]^p$).

The validity of the statement of Ramsey's Theorem for a particular choice of k,m,p is denoted by $n \longrightarrow (m)^p_k$ (the Erdős-Rado *partition arrow*), $r(m,p,k)$ is the *Ramsey number* which is the smallest n with $n \longrightarrow (m)^p_k$.

The above statement may then be formulated as follows:

$$\text{FRT: } (\forall k,p,m)(\exists n)(n \longrightarrow (m)^p_k).$$

There exists an infinite version of this theorem which states

$$\text{IRT: } (\forall k,p)(\omega \longrightarrow (\omega)^p_k)$$

and FRT follows from IRT by a standard compactness argument. (The statement of FRT is equivalent to the fact that the chromatic number of the set system $(\binom{X}{p},\{\binom{Y}{p};\ Y\in\binom{X}{m}\})$ is at least $k+1$ so we may use [1] directly).

Erdős and Rado [13],[14] generalized IRT to all cardinals. However, the line of research inspired by the Transfinite Ramsey Theorem is not covered in this paper (although some of the topics covered here were partially motivated by tranfinite methods, see §§4,8 below).

II. Schur's theorem [10]: For every k there exists an n such that for every partition of the set $[1,n] = \{1,2,\ldots,n\}$ into k classes there are numbers $i,j \leq n$ such that all the numbers $i, j, i+j$ belong to one of the classes of the partition (one may take $n = [k!e]$).

This is the earliest theorem of the type we are going to discuss here. It is an easy consequence of FRT and so is the following strengthening of it.

Rado's theorem [8] For every k,m there exists an n such that for every partition of the set $[1,n]$ into k classes there are numbers $i_1,\ldots,i_m \leq n$ such that all sums of the form $\sum_{j \in x} i_j$ where $\emptyset \neq x \subseteq [1,m]$ belong to one of the classes. In §§5 and 9 we shall prove stronger versions of these two theorems.

III. Van der Waerden's theorem [12]: For all k,m there exists a natural number n with the following property: for every partition of $[1,n]$ into k classes there exists an arithmetic progression of length m in one of the classes.

Van der Waerden's theorem is the starting point of many generalizations and variations (both combinatorial and number theoretical). Only a very small part of it will be reviewed below.

The above examples are canonical examples of theorems of *Partition Theory* and they are often described as *Ramsey Type statements*. Roughly speaking a Ramsey type statement relates two things:

combinatorial partitions (i.e. colourings)

and

a particular structural property (such as sums, power sets, etc.)

One may approach Ramsey type statements from different angles depending on the particular questions one is trying to answer. The approach taken in this paper is axiomatized in §3 using the terminology of category theory. (However, category theory will not be used in most of the paper). For the approach taken in this paper the following are typical problems.

Ramsey graph problem: Which graphs G have the property that for every partition of the edges of G into two classes at least one of the classes contains a triangle?

Schur set problem: Which sets S of positive integers have the property that for every partition of S into a finite number of classes one of the classes contains two numbers together with their sum?

Van der Waerden problem: Which sets V of positive integers have the property that for every partition of V into a finite number of classes one of them contains an arithmetic progression of a given length?

Many variants of these questions were asked by P. Erdős who has been the main force behind the development of partition theory.

In their full generality the above problems seem to be intractable. But they led to many interesting particular problems and some beautiful theorems. Szemerédi [11], Deuber [4], Graham-Leeb-Rothschild [6] gave such examples. Because of their complexity these examples are not covered here. Most of the material presented here concerns graphs and set systems and typical examples are stated below in §2 (I-V).

Purely combinatorial statements present a smaller part of this paper. However several new results are presented (mainly in §2 but they are also scattered throughout the text). Certain areas of partition theory are not covered here at all (e.g Ramsey minimal graphs, ordering properties, Ramsey numbers and their generalizations). The reason is that these questions are covered elsewhere (see e.g. [2],[3],[5]) and that they do not fit into the framework of this paper. As a whole, [7] presents a more exhaustive list of recent theorems.

Here, the main attention is devoted to applications and, more generally, to connections of partition theory with other branches of mathematics. The authors are convinced that these connections are at present probably the most interesting aspect of partition theory. These connections have a wide spectrum and of course only a smaller (and less important) part of the research was carried out by the authors. Proper credit for a particular result is given at the end of this article.

Let us add a few remarks concerning the terminology and notation.

Natural numbers are denoted by $i,j,k,m,n,\ldots,$ the set of all natural numbers by $\mathbb{N}$. Ordinals are denoted by $\alpha,\beta,\ldots$. An ordinal is sometimes used to denote the set of all smaller ordinals (i.e. $k = \{0,1,\ldots,k-1\}$, $\omega = \mathbb{N}$). All graph theoretical and combinatorial notions have their standard meaning and are mostly defined at appropriate places. The different parts of the paper are more or less independent of each other.

A word concerning "partition terminology". The final part of the statement of FRT (see above) "for every partition of the set $\binom{X}{p}$ into k classes there exists an m-element

subset $Y \subseteq X$ all of whose p-element subsets belong to one of the classes of the given partition"
will be rephrased as
"for every colouring of the set $\binom{X}{p}$ by k colours there exist a homogeneous m-element set"
or
"for every colouring $c: \binom{X}{p} \longrightarrow k$ there exists a homogeneous m-element set".

Further combinations of these notions are possible and it is hoped that they will never lead to confusion.

Most of the paper concerns finite sets although infinite sets are mentioned from time to time. For most questions discussed in this paper little is known for infinite sets. However even some finite partition theorems have some infinite applications. Somewhat surprisingly the distinction between finite and infinite methods is sometimes very subtle.

§2 Sample results

In this part we state some of the results which will either be needed later on or represent a new development. To keep this survey readable we do not formulate the theorems in their strongest forms.

I. Set systems of a given type

Definition 2.1: A type Δ is a finite sequence of positive integers $(d_i;\ i \in I)$. An *ordered set system of type* Δ is a pair $(X, \mathcal{M})$ where X is a finite ordered set and $\mathcal{M} = (\mathcal{M}_i;\ i \in I)$ where $\mathcal{M}_i \subseteq \binom{X}{d_i}$.

$(X, \mathcal{M})$ is an (induced) *subsystem* of $(Y, \mathcal{N})$, $\mathcal{N} = (\mathcal{N}_i;\ i \in I)$ if

$$X \subseteq Y \quad \text{and} \quad \mathcal{N}_i \cap \binom{X}{\delta_i} = \mathcal{M}_i \quad \text{for every } i \in I .$$

By an *isomorphism* of ordered set systems we mean a monotonic isomorphism.

Theorem 2.2: For every positive integer k and every ordered set systems F, X of type Δ there exists an ordered set system Y of type Δ with the following property: for every partition of all subsystems of Y which are isomorphic to F into k classes there exists a subsystem X' of Y, X' isomorphic to X such that all the subsystems of X' which are isomorphic to F belong to one of the classes.

Moreover, if X does not contain an irreducible set system L then Y may be chosen with the same property.

Here: $L = (Z,P)$, $P = (P_i;\ i \in I)$, is called irreducible iff for every two distinct vertices $x,y \in Z$ there exists $M \in \bigcup_{i \in I} P_i$ such that $\{x,y\} \subseteq M$.

Corollary 2.3: Let (X,M) be a k-graph (i.e. $M \subseteq \binom{X}{k}$). Then there exists a k-graph (Y,N) such that for every partition of $\binom{Y}{k}$ into two classes there exists a subgraph (X',M') of (Y,M) such that each of the sets $\binom{X'}{k} \cap M'$ and $\binom{X'}{k} - M'$ belongs to one of the classes of the partition.

These results were proved in [9]. The first part of Theorem 2.2 was independently proved in [1]. In terms of §3 below Theorem 2.2. asserts that the class of all ordered set systems of a given type (eventually without an irreducible subsystem) is a Ramsey category. Actually [9] contains more technical statements.

Originally the proof of this statement was not an easy one. However the proof of Corollary 2.3 may be simplified using the following lemma (discovered independently by Bruce Rothschild, see [13]):

Given positive integers k and a, a k-graph $(X,\mathcal{M})$ is said to be *a-partite* if there exists a partition $X = X_1 \cup \ldots \cup X_a$ such that $|M \cap X_i| = 1$ for every $M \in \mathcal{M}$ (symbolically we write $\mathcal{M} \subset X_1 \times \ldots \times X_a$). An a-partite graph $((X_1,\ldots,X_a),\mathcal{M})$ is a subgraph of $((Y_1,\ldots,Y_a),\mathcal{N})$ if $X_i \subseteq Y_i$ and $(\bigcup X_i,\mathcal{M})$ is an (induced) subsystem of $(\bigcup Y_i,\mathcal{N})$.

<u>a-partite lemma 2.4</u>: For every positive integer k and for every a-partite graph $(X,\mathcal{M})$ there exists an a-partite graph $(Y,\mathcal{N})$ such that for every partition of $\mathcal{N}$ into k classes there exists an a-partite graph $(X',\mathcal{M}')$ isomorphic to $(X,\mathcal{M})$ such that $\mathcal{M}'$ belongs to one of the classes.

After proving the a-partite lemma (which is a consequence of Ramsey's theorem), Corollary 2.3 follows relatively easily by downward induction on the number a.

II. <u>Sparse graphs</u>

Here and elsewhere in §2, we consider (for simplicity) graphs only.

Suppose that $G = (V,E)$ is a Ramsey graph for K_k, the complete graph with k vertices, i.e. we assume that for every partition of E into 2 classes one of the classes contains a subgraph isomorphic to K_k. A set E', $E' \subseteq E$ such that $(\bigcup E',E') \simeq K_k$ is called a *copy* of K_k. Denote by $(E,\mathcal{E})$ the $\binom{k}{2}$-graph $(E,\{E';E'$ is a copy of $K_k\})$. The fact that the chromatic number of $(E,\mathcal{E})$ is > 2 is equivalent to the fact that G is Ramsey for K_k.

In this connection the following natural question arises: how dense has this set system got to be? [4]. Partial results related to this question were obtained in [14]. The following gives a full answer to this question:

Theorem 2.5: For every k and every $r \geq 2$ there exists a Ramsey graph $G = (V,E)$ such that the set system $(E,\mathcal{E})$ does not contain cycles of lengths less than r.

Corollary 2.6: For every k there exists a Ramsey graph G such that any two copies of K_k intersect in at most one edge.

These results may be generalized from complete graphs to a wider class of graphs.

III. Forbidden subgraphs

Let $\mathcal{Q}$ be a class (finite or infinite) of graphs. Denote by Forb($\mathcal{Q}$) the class of all finite graphs G such that A fails to be an (induced) subgraph of G for every $A \in \mathcal{Q}$.

We say that Forb($\mathcal{Q}$) has the *edge-partition-property* (see §3 for the general definition of partition property) if for every graph $G \in$ Forb($\mathcal{Q}$) and for every positive integer k there exists a graph $H \in$ Forb($\mathcal{Q}$) such that for every partition of the edges of H into k classes there exists a homogeneous copy of G in H.

Edge-partition properties of classes Forb($\mathcal{Q}$) were studied in [7],[8],[9] and they are not easy to establish. By a refinement of our methods we are able to prove the following which generalizes all previously known results concerned with classes Forb($\mathcal{Q}$).

Definition 2.7: We say that G is a k-*chromatically connected* graph if every graph induced on a cut set of G has chromatic number $\geq k$.

Obviously for every triangulated (i.e. rigid circuit) graph G the connectivity of G and the chromatic connectivity coincide.

Theorem 2.8: Let $\mathcal{Q}$ be a set of 3-chromatically connected graphs. Then the class Forb($\mathcal{Q}$) has the edge-partition property.

Several classes of triangulated graphs are described in [2]. Observe also that the class of all 3-chromatically connected graphs is closed under products.

IV. Classes of graphs defined by homomorphisms

Given two graphs $G = (V,E)$, $G' = (V',E')$ a mapping $f: V \longrightarrow V'$ is called a *homomorphism* if $\{x,y\} \in E$ implies $\{f(x),f(y)\} \in E'$ Denote by $\langle G,G'\rangle$ the set of all homomorphisms $G \longrightarrow G'$.

The following are basic classes defined by homomorphisms

$$G \longrightarrow \; = \{H;\ \langle G,H\rangle \neq \emptyset\},$$
$$G \not\longrightarrow \; = \{H;\ \langle G,H\rangle = \emptyset\},$$
$$\longrightarrow G = \{H;\ \langle H,G\rangle \neq \emptyset\},$$
$$\not\longrightarrow G = \{H;\ \langle H,G\rangle = \emptyset\},$$

where G is a fixed graph (further classes may be defined from these graphs by set operations). These classes were studied from the category theory point of view in [5]. Here we discuss edge-partition properties of these classes.

The classes $G \not\longrightarrow$ were discussed above in III. Using the fact that any homomorphic image of a 3-chromatically connected graph is 3-chromatically connected we get, in particular, that $G \not\longrightarrow$ has the edge-partition property for every 3-chromatically connected graph.

The classes $G \longrightarrow$ and $\not\longrightarrow G$ always have the edge-partition property while the class $\not\longrightarrow G$ has the edge-partition property iff G is a bipartite graph. (The last statement reflects that fact that $\chi(H) > \chi(G)$ whenever H is Ramsey for a non-bipartite G; this question is discussed in [3], see also §8 for a related question for selective graphs). Considering these remarks it seems that the following modification of classes $\longrightarrow G$ is a more appropriate notion in the present context.

Definition 2.9: Let p be a positive integer, and G be a fixed graph. Denote by $\xrightarrow[p]{} G$ the class of all graphs H with the property that for any subgraph H' of H with at most p vertices there exists a homomorphism $H' \longrightarrow G$.

Examples: $\xrightarrow[100]{} K_2$ is the class of all graphs without odd cycles of lengths < 100;

$\xrightarrow[100]{} C_5$ is a class of all graphs which can be "locally" mapped into the pentagon;

$\xrightarrow[p]{} K_k$ is a class of all "locally" k-colourable graphs.

We say that a graph G is *reduced* if every homomorphism $G \longrightarrow G$ is an isomorphism. It is a simple observation that for every graph G there exists a uniquely determined reduced subgraph G' of G for which there exists a homomorphism $G \longrightarrow G'$ (this is explicitly shown in [5]). Moreover, it follows that $\xrightarrow[p]{} G = \xrightarrow[p]{} G'$.

Using these definitions we may prove the following

Theorem 2.10: Let G be a reduced graph. The following two statements are equivalent:

(1) For every positive integer p the class $\xrightarrow[p]{} G$ has the edge-partition property;

(2) G is edge transtive.

The necessity of edge transitivity is clear as if G fails to be an edge transitive graph then for a suitable amalgamation of two copies of G there exists no homomorphism into G. Proof of the sufficiency is much more difficult and proceeds by downward induction on $|G|$ and p (the main result of [8] is the starting case: $K_{\chi(G)}$ is the minimal homomorphic image of the graph G).

In particular, the class Cyc_{2k+1} ($= \xrightarrow[2k+1]{} K_2$) of all graphs without cycles of odd length $\leq 2k+1$ has the edge partition property ([12]).

V. Clique subgraphs and metric subgraphs

In the terminology of §3 the above theorems assert edge-partition properties of various categories of graphs and all embeddings. However embeddings are not the only interesting mappings in this context. Several other notions were studied and we mention in this section two stronger results

Definition 2.11: Let $G = (V,E)$, $G' = (V',E')$ be graphs, G an (induced) subgraph of G' (i.e. $V \subset V'$ and $E = \{e \in E';\ e \subseteq V\}$). We say that G is a *clique subgraph* of G' if every clique of G is a clique of G'. We say that G is a *metric subgraph* of G' if the metric of G and the metric of G' restricted to V coincide (i.e. $d_G(x,y) = d_{G'}(x,y)$ for all $x,y \in V$).

Theorem 2.12: For every graph G and for every positive integer k there exists a graph H such that for every partition of cliques of H into k classes there exists a clique subgraph G' of H isomorphic to G such that the colour of a clique in G' depends on its size only.

Theorem 2.13: For every graph $G = (V,E)$ and every positive integer k there exists a graph $H = (W,F)$ such that for every partition $[W]^2 = a_1 \cup \ldots \cup a_k$ there exists a metric subgraph G' of H, G' is isomorphic to G, such that the colour of $\{x,y\}$ depends only on $d_{G'}(x,y) (= d_H(x,y))$.

These two theorems also generalize some of the theorems originally proved in [6],[8],[9].

There is more in the last theorem than meets the eye. This will be briefly sketched below:

An easy variant of the edge-partition property of the class of all bipartite graphs yields the following:

<u>Corollary 2.14</u>: For every matrix $A = (a_{ij})_{i=1,j=1}^{m,n}$ and every positive integer k there exists a matrix $B = (b_{ij})_{i=1,j=1}^{M,N}$ such that for every colouring of $[1,M]\times[1,N]$ by k-colours there exists a submatrix $A' = (a'_{r(i),s(j)})_{i=1,j=1}^{m,\ n}$ such that $a'_{r(i),s(j)} = a_{ij}$ and the colour of $(r(i),s(j))$ depends on $a'_{r(i),s(j)}$ only. (In other words; for every colouring of entries of B there exists a submatrix A' of B isomorphic to A such that the colour of any entry of A' depends only on its value).

This may be strengthened to the following which is a consequence of Theorem 2.2 above for the case $\Delta = (2_1 2_1 \ldots 2)$:

<u>Corollary 2.15</u>: For every symmetric matrix $A = (a_{ij})_{i,j=1}^{n}$ and for every positive integer k there exists a symmetric matrix $B = (b_{ij})_{i,j=1}^{N}$ such that for every symmetric k-colouring of $[1,N]\times[1,N]$ there exists a submatrix $A' = (a'_{r(i),r(j)})_{i,j=1}^{n}$ isomorphic to A such that the colour of an entry of A' depends only on its value. By symmetric k-colouring we mean that the colour of (i,j) and (j,i) coincide.

This is further strengthened by Theorem 2.13.

Corollary 2.16: For every distance matrix $A = (a_{ij})_{i,j=1}^{n}$ of a graph there exists a distance matrix $B = (b_{ij})_{i,j=1}^{N}$ of a graph such that for every symmetric k-colouring of entries of B there exists a submatrix A' of B isomorphic to A such that the colour of an entry of A' depends only on its value.

It would be interesting to extend these theorems to further classes of finite integer-valued matrices. For example the following problem is of particular interest:

We say that a matrix $A = (a_{ij})_{i,j=1}^{n}$ is a *partial latin square* (PLS) if a_{ij} is a natural number for every i,j;

$$0 \neq a_{ij} = a_{ij'} \implies j = j'$$

$$0 \neq a_{ij} = a_{i'j} \implies i \neq i'.$$

A is isomorphic to $A' = (a'_{ij})_{i,j=1}^{n}$ iff

$$a_{ij} = 0 \iff a'_{ij} = 0$$

$$a_{ij} = a_{i'j'} \iff a'_{ij} = a'_{i'j'} .$$

Problem 2.17: Let A be a PLS, k positive integer. Is it true that there exists a PLS B such that for every k-colouring of entries of B there exists a submatrix A' of B, A' isomorphic to A, such that the colour of an entry of A' depends only on its value?

If true then this statement implies the edge partition property of the class of all 3-graphs without 2-cycles (a k-graph is a pair $(X,\mathcal{M})$ where $\mathcal{M} \subseteq \binom{X}{k}$; $(X,\mathcal{M})$ doesn't contain 2-cycles if $|M \cap M'| \leq 1$ for every two different edges M,M' of $\mathcal{M}$). This in turn would be the first step in solving a major problem in partition theory:

Problem 2.18: Is it true that the class of all graphs without triangles and rectangles has the edge-partition property?

This problem is related to the edge-partition property of k-graphs without 2-cycles (and consequently to Problem 2.17 above) by the following statement:

Proposition 2.19: Assume that for every k the class of all k-graphs without 2-cycles has the edge-partition property. Then the class of all bipartite graphs without rectangles has the edge-partition property.

We close this section by a sketch proof of Proposition 2.19:

Proof: Let $G = (X,X',E)$ be a bipartite graph. Assume, without loss of generality, that $d_G(x) = r$ for every $x \in X'$. Define the r-graph $(X,\mathcal{M})$ as follows:

$M \in \mathcal{M}$ iff $M = \{y \in X; [x,y] \in E\}$ for some x. $(X,\mathcal{M})$ does not contain 2-cycles. Put $K = k(r-1)+1$. Let $(Y,\mathcal{N})$, $Y = \{y_1,\ldots,y_n\}$ be a K-graph without 2-cycles with the following property: if κ is a k-element subset of $\{1,2,\ldots,k(r-1)+1\}$ and if we consider the k-graph $(Y,\mathcal{N}|\kappa) = (Y,\{N|\kappa; N \in \mathcal{N}\})$ then $(Y,\mathcal{N}|\kappa)$ contains $(X,\mathcal{M})$ as an induced subgraph.

(Here we put $N|\kappa = \{x_i; i \in \kappa\}$ for $N = \{x_1,\ldots,x_K\}$; $(Y,\mathcal{N})$ can easily be constructed if we blow up $(X,\mathcal{M})$ and take disjoint unions.) Now by the assumed edge-partition property of the class of all K-graphs without 2-cycles there exists a K-graph $(Z,\mathcal{P})$ without 2-cycles such that for every k-colouring of $\mathcal{P}$ there exists a homogeneous copy $(Y',\mathcal{N}')$ of $(Y,\mathcal{N})$. Moreover, using the ordering property (see [10],[11]) of the classes without 2-cycles we may assume that the copy

$(Y',\mathcal{N}')$ is monotone isomorphic to $(Y,\mathcal{N})$. Define the bipartite graph $H = (Z,\mathcal{P},E)$ by putting $E = \{[z,P];\ z\in P\}$. Obviously H does not contain rectangles. One can also prove that H is a Ramsey graph for G.

§3 Applications to category theory

It has been observed by K. Leeb that certain areas of partition theory find a suitable formulation in terms of the theory of categories. This is emphasized in [2],[5],[3],[17],[7],[10],[13]. One proceeds as follows:

Given a category K and its object A denote by $\binom{\ }{A}$ the following functor from K into the category set (of all sets and all mappings).

For an object B of K we define

$$\binom{B}{A} = \{[f];\ f \text{ is a morphism } A \longrightarrow B \text{ in } K\}$$

where $[f]$ is the class of the equivalence $\sim$ (containing f) generated by $f \sim g$ if $f = g\circ h$ for an isomorphism $h: A \longrightarrow A$ (in K);

For a K-morphism $f: B \longrightarrow C$ we define $\binom{f}{A}:\binom{B}{A} \longrightarrow \binom{C}{A}$ by

$$\binom{f}{A}([g]) = [f\circ g] \quad \text{for every } g \in \binom{B}{A}.$$

Given objects A,B,C and a cardinal α we say that C is an (A,α)-Ramsey for B if for every colouring $c: \binom{C}{A} \longrightarrow \alpha$ there exists a K-morphism $f: B \longrightarrow C$ such that $c\circ\binom{f}{A}$ is a constant mapping. This fact is denoted by various symbols mostly inspired by the Erdős-Rado partition arrow:

$$C \longrightarrow (B)^A_\alpha,\quad C \xrightarrow[\alpha]{A} B,\quad C \xrightarrow[\alpha]{} (B)^\alpha,\quad \text{or}\quad C \longleftrightarrow (B)^A_\alpha.$$

The choice depends on the context in which a particular symbol is used. The motivation for our notation lies in the fact that the arrows $\xrightarrow[\alpha]{A}$ form a category $(B \xrightarrow[\alpha]{A} B' \xrightarrow[\beta]{A} C$ implies $B \xrightarrow[\alpha\beta]{A} C)$ and they have the intuitive meaning of "combinatorially strengthened" morphisms of K.

In this survey we shall omit the partition symbols whenever possible.

We say that a category K has the A-*partition property* if for every object $B \in K$ and every positive integer k there exists an (A,k)-Ramsey object for B (i.e. an object C such that $B \xrightarrow[k]{A} C$ in K). In the extremal case we say that K is a *Ramsey category* if K has the A-partition property for every object A.

The notion of a Ramsey category is very strong and consequently not many Ramsey categories are known. Essentially, the following is the complete list of all non-trivial Ramsey categories:

-Δ = the simplicial category = the category of all finite sets and all 1-1 mappings, [15];

-$Vect_Q$ = the category of all finite vector spaces (over a fixed field Q), [2];

-$Afin_Q$ = the category of all finite affine spaces (over a fixed field Q), [2];

-[A] = the Hales-Jewett category, [1],[5];
This is a suitable categorical extension of the Hales-Jewett theorem [4]. As follows from the recent simplification of [2] by Spencer [16], this theorem is the heart of the above three theorems;

-Deub[A] = the Deuber category (of suitable finite subsets of integers; this is related to an axiomatization of so-called Rado-sets of integers which in turn yields the solution of a conjecture of Rado [14]), see [1],[5,6];

- $\overrightarrow{\mathrm{Soc}}(\Delta)$ = the category of all ordered finite set systems of type Δ, see [10];

- $\overrightarrow{\mathrm{Forb}}_{\Delta}(\mathcal{Q})$ = the category of ordered finite set systems of type Δ, which do not contain a subsystem belonging to the set $\mathcal{Q}$ (see Theorem 2.2 above for a precise statement of this result), see [9,10].

Several other (more technical) examples of Ramsey categories are listed in [13]. Moreover, [5] contains a categorical procedure for generating Ramsey categories: If Tree ($\mathcal{K}$) denotes the category of all finite trees over a category $\mathcal{K}$ (i.e. trees T with object of $\mathcal{K}$ substituted for nodes of T) then $\mathcal{K}$ Ramsey implies Tree ($\mathcal{K}$) Ramsey (the precise statement of this result may be found in [5] and [3]).

However, many common categories fail to be Ramsey for a trivial reason. For example, the category Gra of all (unordered) graphs fails to be Ramsey as there exists no $(P_2,2)$-Ramsey graph for the pentagon C_5. (P_2 is the path of length 2). For suppose that $G = (V,E)$ is a $(P_2,2)$-Ramsey graph for C_5. Let $\leq$ be a (total) orfering of V and define a colouring $\binom{G}{P_2} = A_1 \cup A_2$ as follows: A_1 is formed by all induced paths of length 2 which are monotone (with the relativized ordering $\leq$) isomorphic to $(\{0,1,2\}, \{\{0,1\},\{1,2\}\})$ and A_2 is the set of all other induced paths of length 2 in G. It is easy to see that any ordering of C_5 contains paths of length 2 of both colours.

With respect to these comments the following seems to be the basic problem of partition theory:

Ramsey problem

For a given category $\mathcal{K}$ characterize the class of all those objects A for which $\mathcal{K}$ has the A-partition property.

A theorem which solves the Ramsey problem for a particular category K is called *a prototype theorem* for K, see [7], and the class of all A for which K has the A-partition property is denoted by $r(K)$.

Thus K is Ramsey iff $r(K) = K$.

The Ramsey problem was solved for several categories. Let us mention at least two results:

$r(\text{Gra})$ is the class of all complete and discrete graphs, see [8]; $r(K)$ is determined for many categories of graphs and set systems, see [10],[13]:

$r(\text{Ab})$ is the class of all groups of the form $\oplus(\mathbb{Z}_{p^{\alpha_p}})^{n_p}$ where for each p, $\min\{\alpha_p, n_p\} = 1$, see [17] (Ab is the category of all finite abelian groups).

A solution of the Ramsey problem is related in many cases to the so-called *ordering property* of a category, see [13]. In [11] the authors proved the ordering property for "nearly all" categories of graphs and set systems. This in turn means that $r(K)$, for a category of set systems, is always a subclass of the class of all set systems which belong to K and which admit any permutation of its vertices for an automorphism. It is interesting to note that $r(K)$ may be a proper subclass of this class as follows from [12]: $r(\text{Cyc}_{2k+1}) = \{K_1, K_2\}$; see §2.

§4 Applications to logic

The infinite Ramsey theorem (here we mean $\omega \longrightarrow (\omega)^p_k$) has been of interest to logicians since its discovery. The reason for this lies in Ramsey's original paper and, more generally, in the fact that Ramsey's theorem presents a powerful statement which (because of its generality) can be used in various situations. In this sense Ramsey's theorem is a logical combinatorial principle whose provability has to be discussed; see e.g. [5]. However only recently have there appeared applications to logic which fit easily into this paper (i.e. finite partition theorems and their applications). In this part we mention two examples of this. They show that the flow of ideas went in both directions. First, we briefly mention a (rather technical) example of an application of Theorem 2.2 stated in §2 above to the construction of special models of Peano Arithmetic. Secondly, we survey the recent spectacular development of partition theory related to a particular version of the finite Ramsey theorem which is true and yet unprovable by finitary means. We sketch a few ideas which stem from logic and which could be of interest to combinatorial mathematicians (for whom this text is intended).

Both examples are concerned with Peano Arithmetic (PA) which is a first-order theory with non-logical symbols $\leq, +, \cdot, s$ (for successor) whose axioms express the common properties of natural numbers and include the induction scheme for formulae

$$\phi(o) \ \& \ (\forall x)(\phi(x) \longrightarrow \phi(s(x))) \longrightarrow (\forall x)(\phi(x));$$

see [14] and any non-combinatorial reference for this section.

Let $\mathcal{Q} = (A, \leq ; +, \cdot, s)$ be a model of PA. A set $\{b_i ; i = 1,2,\ldots\}$, $b_1 < b_2 < \ldots$, is called a set of *indiscernibles* if for any choice $\vec{b}' = b_{i(1)} < \ldots < b_{i(k)}$, $\vec{b}'' = b_{j(1)} < \ldots < b_{j(k)}$ and any formula ϕ with k free variables $\phi(\vec{b}')$ is valid in $\mathcal{Q}$ iff $\phi(\vec{b}'')$ is valid in $\mathcal{Q}$.

It follows from the Erdős-Rado transfinite extension of Ramsey's theorem (see [4]) that large models of PA (more precisely: models of size α where $\alpha \longrightarrow (\omega)^{<\omega}$) contain an infinite set of indiscernible elements (essentially we define a colouring according to the validity of $\phi(\vec{a},\vec{b})$. [1] is devoted to the quantitative study of this fact and the sizes of models of PA without an infinite set of indiscernible elements are completely characterized by the Ramsey function (for infinite cardinals). However in order to carry out the necessary constructions of models of PA without an infinite set of indiscernibles a delicate question in combinatorics arose and this led to an independent discovery of Theorem 2.2 above in the following form:

A system of colours of length n, $\alpha = (\alpha_o,\ldots,\alpha_n)$ is an $n+1$ sequence of finite, non-empty sets. An α-coloured set consists of a finite ordered set X and a function $f\colon [X]^{\leq n} \longrightarrow \alpha_o \cup \alpha_1 \cup \ldots \cup \alpha_n$ such that $f(a_o,\ldots,a_{k-1}) \in \alpha_k$. $f(A)$ is called the colour of A. An α-pattern is an α-coloured set whose underlying ordered set is an integer (called the length of the pattern). Each α-coloured set is monotonic isomorphic to the unique pattern, called the α-pattern (of α-coloured set).

Of course, each α-coloured set (X,f), $f\colon [X]^{\leq n} \longrightarrow \alpha_o \cup \ldots \cup \alpha_n$ corresponds to a set system $(X,(M_j;\ j \in \bigsqcup_{i=o}^{n} \alpha_i))$ defined by $M \in M_j$ iff $f(M) = j$ (here $\bigsqcup$ stands for the disjoint union).

A repeated use of Theorem 2.2 for every different α-pattern of length p implies then the following:

<u>Theorem 4.1</u> [1]: For all natural numbers n,p,k, a system α of colours of length n and an α-pattern P there exists

an α-pattern Q such that for any α-coloured set (X,f) with α-pattern Q and for any colouring $c: [X]^p \longrightarrow k$ there exists $Y \subseteq X$ such that Y has α-pattern P under (the restriction of) f and for $A \in [X]^p$, $c(A)$ depends only on the α-pattern of A under f.

The second recent connection between logic and partition theory drew a lot of attention among logicians and as it led to several new combinatorial ideas we state it in some detail.

All sets considered below in this part are subsets of natural numbers and the set $\{0,1,\dots,n-1\}$ will be denoted by n.

A finite set A is called *relatively large* if $\min A \le |A|$. Using this notion we have the following strengthening of the finite Ramsey theorem [12]:

For all natural numbers m,k,p there exists an n with the following property: for every colouring $c: [n]^p \longrightarrow k$ there exists a set $A \subseteq n$, $|A| \ge m$ such that c restricted to the set $[A]^p$ is a constant mapping.

This statement is denoted by $n \underset{*}{\longrightarrow} (m)^p_k$ (proof of this statement is simple: Assume contrary for a particular choice of m,p,k. Hence for every n there exists a counterexample c_n. Using a compactness argument (or König's infinite tree lemma) there exists a $c:[\mathbb{N}]^p \longrightarrow k$ such that for every n the restriction $c|_{[n]^p}$ is a counterexample again. But the infinite Ramsey theorem yields an infinite set X such that $c|_{[X]^p}$ is a constant. Putting $X = \{x_1,x_2,\dots\}$ we get a homogeneous set $\{x_1,x_2,\dots,x_{x_1}\}$ which is relatively large, a contradiction).

This is the easier part of [12]. The main part consists in showing that one cannot prove the above statement by means of finite sets only. More precisely: Denote by ϕ the following formula:

$$(\forall m,k,p)(\exists n)(n \underset{*}{\longrightarrow} (m)^p_k).$$

ϕ is a formula about finite sets. But ϕ can be regarded as a formula of PA, as finite sequences and consequently all finite set notions may be coded by natural numbers (e.g. the sequence $x(0),x(1),\ldots,x(k)$ may be coded by the numbers $p_o^{x(0)+1} p_1^{x(1)+1} \ldots p_k^{x(k)+1}$, where p_k is the k-th prime ≥ 2); the existence of such encoding is the heart of Gödel's First Incompleteness Theorem. We say that ϕ is provable in PA if the encoded formula of ϕ is provable in PA.

Theorem 4.2 [12]: 1) ϕ is a true statement (i.e. it is valid in the standard model $\mathbb{N}$ of PA);

2) ϕ cannot be proved within PA.

In other words: ϕ is an interesting example of a formula which demonstrates the incompleteness of the PA, a concrete example which demonstrates Gödel's First Incompleteness Theorem.

Several other combinatorial statements were proved to be undecidable in PA, see [10],[11]. All these statements are examples of Ramsey-type statements, though they need not be. In [13] there is exhibited an example of a combinatorial statement undecidable in PA which is not of the Ramsey type.

Several proofs of Theorem 4.2 are by now available. The original proofs were model theoretic and (for a non-specialist) gave little information about the reason "why ϕ is unprovable". Recently, however, different proofs were given in

[16] and [6]. These proofs are, from the combinatorial side, even more interesting. They depend on the following ideas:

First, denote

$$r_*(k,p,m) = \min\{n;\ n \underset{*}{\longrightarrow} (m)_k^p\}.$$

It is difficult to obtain tight bounds for (standard) Ramsey numbers $r(k,p,m)$. But for numbers $r_*(k,p,m)$ it is difficult to obtain *any* kind of bounds. Particular bounds were given in [3] for the function $r_*(k) = r_*(2,2,k)$. (In fact [3] considers the function

$$R_*(k) = \min\{n;\ [k,n] \longrightarrow (3)_2^2\}.$$

Then $R_*(1) = 6, R_*(2) = 8$, $R_*(3) = 13$ and for this simplest case relatively tight bounds were obtained

$$2^{2^{\alpha.k}} < R_*(k) < 2^{2^{3k \log k}} , \quad \alpha < \frac{1}{2}).$$

The astronomical rate of growth of the function $r_*(k,p,m)$ was demonstrated by R. Solovay in [16] by means of the diagonal function $\sigma(n) = r_*(n,n,n+1)$. It is proved in [16] by combinatorial means that if f is any recursive function which can be proved in PA to be total (i.e. defined for all natural numbers) then $f(n) < \sigma(n)$ for all sufficiently large n. (In particular, if σ is recursive, it follows immediately that we cannot prove in PA that σ is a total function). This provides an alternative proof of [12] and establishes a general lower bound for *-Ramsey numbers. The proof is based on a suitable hierarchy of recursive functions and on careful partition combinatorics. (An interesting terminology is suggested: a pair (X,F) is called a (b,c)-algebra if $F\colon [X]^b \longrightarrow c$).

The combinatorial side of *-Ramsey numbers was pursued further by J. Ketonen in [6]. Here the problem is essentially restated and, in a way, the problem of *-Ramsey numbers solved. The main idea of [6] is the following:
Let α be a countable ordinal. If α is a limit ordinal choose an increasing sequence $\{\alpha\}$ converging to α (the terms of this sequence are denoted by $\{\alpha\}(1), \{\alpha\}(2), \ldots$).
If $\alpha = \beta+1$ put $\{\alpha\}(n) = \beta$.
Let ε_o be the limit of the sequence $\omega, \omega^{\omega}, \omega^{\omega^{\omega}}, \ldots$. For all $\alpha \leq \varepsilon_o$ one can give a simple description of such $\{\alpha\}$.
The following is the principal definition of [6]:

<u>Definition 4.3</u>: A set A is 0-large iff $A = \emptyset$.
Let $A = \{a_o < a_1 < \ldots < a_n\}$ be a set of natural numbers.
We say that A is α-large if $A\ \{a_o\}$ is $\{\alpha\}(a_1)$ large.

Clearly A is k-large iff $|A| = k+1$ and if $\{\omega\}(n) = n$ then any ω-large set is relatively large. The finite Ramsey theorem then states the following:
For every k, m, p there exists n such that if X is any n-large set then for every $c: [X]^p \longrightarrow k$ there exists a homogeneous m-large set. Ketonen investigated a similar problem: Determine the ordinals α with the following property. If X is an α-large set then for every $c: [X]^p \longrightarrow k$ there exists a homogeneous ω-large set (i.e. a relatively large homogeneous set). The natural strengthening of the above *-partition property is denoted by $\alpha \xrightarrow[S]{} \beta$; this notation is used in the following theorem of Ketonen.

<u>Theorem 4.4</u> [6] $\qquad \varepsilon_o \xrightarrow[S]{\quad} \alpha$ for every $\alpha < \varepsilon_o$

$\qquad \alpha \not\xrightarrow[S]{\quad} \omega$ for every $\alpha < \varepsilon_o$.

The idea of measuring finite sets by means of ordinals

$< \omega_1$ seems to be fruitful. Another example of this kind is the recent research done on infinite Sperner systems. We sketch it only briefly. It is due to P. Pudlák and the authors and it will appear in full elsewhere.

Let S be a family of finite subsets of an infinite set $A = \{a_1 < a_2 < \ldots\}$ of natural numbers.

Definition 4.5: We say that

1) S is *Ramsey* if for every partition of S into a finite number of classes there exists an infinite set B such that all subsets of B which belong to S are in one of the classes of the partition.

2) S is *Sperner* if there are no $M, N \in S$ such that $M \subsetneq N$.

3) [7] S is *thin* if there are no $M, N \in S$ such that $M \neq N$ and M is a left segment of N.

These notations were analysed by means of the following:

Definition 4.6: Let α be a countable ordinal. We say that S is *α-uniform* if the following holds:

If $\alpha = 0$ then $S = \{\emptyset\}$,

If $\alpha = \beta+1$ then $S_n = \{M-\{a_n\};\ M \in S,\ \min M = a_n\}$ is β-uniform for all n.

If α is limit then S_n is α_n-uniform for a strictly increasing sequence $\alpha_1, \alpha_2, \ldots$ ordinals with $\lim \alpha_n = \alpha$.

Obviously if S is k-uniform then $|M| = k$ for every $M \in S$ and if S is ω-uniform then S contains relatively large sets only. The following holds:

Theorem 4.7: For a family S of finite subsets of natural numbers the following statements are equivalent

1) S is Ramsey;

2) There exists an infinite set A such that S restricted to A is either α-uniform for some α or empty;

3) There exists an infinite set A such that S restricted to A is Sperner;

4) There exists an infinite set A such that S restricted to A is thin;

5) There exists an infinite set A such that S restricted to A does not contain two disjoint uniform families.

This theorem generalizes [7], Theorem 1 (where it is proved 4) $\Longrightarrow$ 1)). As a corollary every α-uniform set system is Ramsey which generalizes the infinite Ramsey theorem.
infinite Ramsey theorem.

Theorem 4.7 has a strong finite version which may be stated as follows:

<u>Theorem 4.8</u>: Let $S = (X, M)$, $X = \{x_1, \ldots, x_m\}$ be a Sperner system, k a positive integer. Then there exists a Sperner system $T = (Y, N)$ such that the following holds: For every partition of N into k classes there exists a subset $X' \subseteq Y$ with the following properties

a) N restricted to X' belongs to one of the classes

b) There exists a partition $X' = \bigcup_{i=1}^{m} X_i$ with the following property: $M' \in N$, $M' \subseteq X'$ iff there exists $M = \{x_{i(1)}, \ldots, x_{i(k)}\}$ such that $M' = \bigcup_{j=1}^{k} X_{i(j)}$.

This is the induced finite version of Theorem 4.7 which is related to Theorem 2.2 (induced subsystems are replaced by induced "blown up" subsystems).

One may also prove selective versions of Theorem 4.7. This is stated in part 8.

§5 Applications to ultrafilters

By an ultrafilter U we mean here a proper non-principal ultrafilter on the set of natural numbers $\mathbb{N}$ (i.e. U is a family of infinite subsets of $\mathbb{N}$ which is closed on finite intersections and which for any set $A \subseteq \mathbb{N}$ contains either A or $\mathbb{N} \setminus A$). The notion of an intrafilter plays an important role in logic, set theory, topology and infinitary combinatorics. [5] is the standard reference text for the theory of ultrafilters; [4] is a recent survey.

There already exist examples of connections between ultrafilters and (essentially) finitary combinatorics. One such example is provided by the following result which was isolated as a useful fact from the theory of ultrafilters:

<u>Lemma 5.1:</u> For every mapping $f: \mathbb{N} \longrightarrow \mathbb{N}$ there exists a partition $\mathbb{N} = A_o \cup A_1 \cup A_2 \cup A_3$ such that f restricted to A_o is the identity and $f(A_i) \cap A_i = \emptyset$ for $i = 1,2,3$.

Clearly this amounts to saying that the graph of any mapping is 3-chromatic (which is explicitly stated already in [3]). There are more recent and less trivial examples. We mention here two examples of connections between partition theory and ultrafilters. They (again!) demonstrate the flow of ideas in both directions (and both of them are related to Fred Galvin).

The first example we want to mention briefly demonstrates a use of ultrafilters for an elegant proof of the following result.

Theorem 5.2 (N. Hindman [8]). Let $\mathbb{N} = A_1 \cup A_2 \cup \ldots \cup A_k$ be a partition. Then there exists an infinite sequence $a_1, a_2, \ldots, a_n$ of natural numbers such that all finite sums of the form $\sum_{j \in \omega} a_{i(j)}$ belong to one of the classes of the partition. Clearly this generalizes Schur's and Rado's theorems (stated in the Introduction). Theorem 5.2 was conjectured in [7]. Hindman's proof is combinatorial and difficult (improved versions are given in [1],[11]). Glaser, following an idea of Galvin, gave an alternative proof of Theorem 5.2 which is based on the following definition:

Definition 5.3: Let U,V be ultrafilters. Denote by $U+V$ the ultrafilter $\{A;\ \{x;\ A-x \in U\} \in V\}$ and by $U.V$ the ultrafilter $\{A;\ \{x;\ A|_x \in U\} \in V\}$ (here $A-x = \{y; x+y \in A\}$ and $A|_x = \{y;\ xy \in A\}$).

Glaser showed that there exists an ultrafilter U_o such that $U_o + U_o = U_o$ which in turn yields a relatively easy proof of Theorem 5.2. The main idea is to define $\mathcal{A}$ as the class of those infinite subsets A of $\mathbb{N}$ which contain an infinite sequence K together with all its finite sums and to prove $U_o \subset \mathcal{A}$. (From this and from the basic property of ultrafilters follows Theorem 5.2). However, if $A \in U_o$ then one may find a desirable sequence $K = \{k_1, k_2, \ldots\}$ inductively by

$$A_o = A,\ k_o \in A_o \cap A_o^*$$

$$A_{i+1} = (A_i - k_i) \cap A_i,\ k_{i+1} \in A_{i+1} \cap A_{i+1}^*,\ k_{i+1} > k_i,$$

where $B^* = \{k;\ B-k \in U_o\}$. The properties of U_o guarantee the correctness of this definition. It is routine to prove that all the finite sums $\sum_{j \in \omega} k_{i(j)}$ belong to A. (A full version of Glaser's proof of Theorem 5.2 is presented in [4]).

Let us remark that the multiplicative version of Theorem 5.2 is also valid as follows either by exponentiation or by the above proof with a multiplicative idempotent $U.U = U$. However the "joint version" of Theorem 5.2 is not valid. This is based on the nonexistence of an ultrafilter U with $U+U = U.U = U$ (announced in [9]).

The second example leads in the opposite direction. The theory of ultrafilters was partially inspired by the Ramsey theorem itself. Intuitively speaking, subsets of $\mathbb{N}$ which belong to an ultrafilter U are "large" sets and every ultrafilter axiomatizes a particular kind of "largeness". On the other hand the simplest case of the Ramsey theorem states that for every partition of pairs we have an infinite homogeneous set. It is natural to ask whether this homogeneous set belongs to a particular ultrafilter U. This leads to the following notions (see [5],[2],[6]): U is a *Ramsey ultrafilter* if for every partition of $[\mathbb{N}]^2$ into 2 classes there exists a homogeneous set belonging to U. This is denoted by $U \longrightarrow (U)^2_2$. More generally U is a *G-arrow ultrafilter* if for every partition of $[\mathbb{N}]^2 = A_1 \cup A_2$ either there exists a set $A \in U$ such that $[A]^2 \subseteq A_1$ or there exists $E \subseteq A_2$ such that the graph $(\cup E, E)$ is isomorphic to G.

From the point of view of the theory of ultrafilters, Ramsey ultrafilters are natural objects of study as there are well-known "less combinatorial" characterizations of them (see [5],[2],[4]).

<u>Theorem 5.4</u>: The following statements are equivalent:

1) U is a Ramsey ultrafilter;
2) $U \longrightarrow (U, K_\omega)$

3) For every partition $\mathbb{N} = \cup A_i$ either one of the classes belongs to U or there exists $A \in U$ such that for every i $|A \cap A_i| \leq 1$ (i.e. U is *selective* [5]; this property gave the name to Ramsey-type theorems which are considered in §8);

4) U is a minimal (non-principal) ultrafilter with respect to the (Rudin-Keisler) ordering $\leq$ defined on the set of all ultrafilters (see [5], chapter 9, for instance, for the definition of $\leq$).

However, it is known by [10] that Ramsey ultrafilters need not exist (in ZFC). Assuming a suitable set-theoretical axiom (such as the continum hypothesis or Martin's Axiom or a "combinatorial principle", see [2]) one can prove the existence.

The notion of G-arrow ultrafilter presents a recent weakening of Ramsey ultrafilters. A question arises to what extent this weakening is effective. One has to be careful as we have the following (providing we extend the definition $U \longrightarrow (U,T)$ to partitions of triples T):

$$U \longrightarrow (U,[4]^3) \text{ iff } U \text{ is Ramsey (see [2]);}$$

[6] contains among other things even

$$U \longrightarrow (U, \text{Fano plane}) \text{ iff } U \text{ is Ramsey.}$$

However for graphs the following is true:

Theorem 5.5 (In the existence parts of this statement below one of the above set-theoretical axioms is assumed):

(1) $U \longrightarrow (U,T)$ for every forest T with finite components only and for every ultrafilter U;

(2) There exists an ultrafilter U such that

$U \longrightarrow (U,G)$ iff G is a forest with finite components only

(3) For every $k \geq 2$ there exists an ultrafilter U such that

$U \longrightarrow (U,K_k)$ and

$U \not\longrightarrow (U,K_{k+1})$

(4) There exists an ultrafilter U such that for every k $U \longrightarrow (U,K_k)$ and $U \not\longrightarrow (U,K_\omega)$ (i.e. U fails to be Ramsey).

((1) is a simple observation made explicit in [12], (2) is proved in [6] and [12], (3) is proved in [6], see also [2] and (4) is proved in [2]).

Proofs of statements of Theorem 5.5 are similar and [2] contains a general technique for proving statements of this type. But these proofs differ in a particular finite partition theorem which is applied: (2) uses the existence of highly chromatic graphs without short cycles (see [14]), (3) uses the existence of Ramsey graphs which do not increase the clique number (see [13] and Theorem 2.2 in §2), (4) uses the assertion $n^k \longrightarrow (n,k+1)^2_2$.

§6 Applications to algebra

In this part we state results concerned with partition properties of lattices and some related questions.

Several (combinatorial) theorems mentioned in this paper may be stated in an algebraical way: For example Ramsey's theorem itself, Ramsey-type theorems for finite vector spaces [6], Schur's theorem and, more generally, Rado-Sanders theorem (see the introduction and [13],[14]) can all be as they represent certain statements about (combinatorial) lattices. Moreover, by an easy combination of Ramsey-type theorems we get theorems which assert directly a partition property of certain algebras. For example Ramsey's theorem and Van der Waerden's theorem together yield the following:

<u>Theorem 6.1</u>: For every finite distributive lattice $(L,\wedge,\vee)$ and every natural number $k \geq 1$ there exists a distributive lattice $(L',\wedge',\vee')$ such that for every colouring $c: L' \longrightarrow \{1,\ldots,k\}$ there exists a sublattice $\bar{L} \subseteq L'$, $\bar{L} \simeq L$ such that c restricted to $\bar{L}$ is a constant.

Using the terminology of §3 this amounts to saying that the category Distr of all finite distributive lattices has the singleton-partition property (or point-partition property).

Other singleton-partition properties of various varieties of lattices were proved in [9] using various constructions. In particular, the singleton-partition property of the class Lat of all finite lattices was established in [9] using a direct construction (which yields a stronger result, see Theorem 6.3 below) and by the use of Pudlák-Tůma's representation theorem of lattices by means of lattices of partitions [11]. However the following is the most general result (a joint result with J. Ježek):

<u>Theorem 6.2</u>: Let $\mathcal{Q}$ be a class of universal algebras (not necessarily a variety) which satisfies

1) $\mathcal{Q}$ is closed on products (i.e. $A,B \in \mathcal{Q} \Longrightarrow A\times B \in \mathcal{Q}$)

2) $\mathcal{Q}$ contains all one-element subalgebras (i.e. if $n \in \mathcal{Q}$ and $x \in |A|$ then x induces a subalgebra of A which belongs to $\mathcal{Q}$).

Then $\mathcal{Q}$ has the singleton-partition property. Explicitly: For every natural number k and every algebra $A \in \mathcal{Q}$ there exists an algebra $B \in \mathcal{Q}$ such that for every partition of elements of B into k classes there exists a subalgebra A' of B, $A' \simeq A$ the elements of A' belonging to one of the classes of the partition.

In particular:

- every variety of idempotent algebras
- every variety of lattices

has the singleton-partition property.

It may be also shown that a similar theorem does not hold for non-idempotent varieties of algebras.

For lattices a stronger result may be proved:

<u>Theorem 6.3</u> [9]: For every natural number k and every finite lattice $(L,\wedge,\vee)$ there exists a finite lattice $(L',\wedge',\vee')$ with the following property: For every partition $c: L' \longrightarrow \{1,\ldots,k\}$ there exists a lattice embedding $f: (L,\wedge,\vee) \longrightarrow (L',\wedge',\vee')$ preserving the covering relation (i.e. a covers b in L implies $f(a)$ covers $f(b)$ in L') such that $c \circ f$ is a constant mapping. Moreover, if the Hasse diagram of L does not contain cycles of length $\leq \ell$ then L' may be chosen with the same property.

Letting L be the 2-element chain we get the existence of a lattice with its Hasse diagram of chromatic number $> k$. This, in response to a problem of Rival was proved independently by Bollobás [1].

The main problem in this area is the question which of the above results may be extended to partitions of subalgebras of a given type rather than just partitions of singletons? As far as partitions of sublattices are concerned we have some negative partition properties. They are based on the ordering property of locally sparse hypergraphs (see [16]). In this sense the above theorems may be considered as a first step towards more difficult theorems. (They represent the first step in the hierarchy

- partitions of vertices,
- partitions of edges (subspaces, subalgebras),
- solution to Ramsey problem.), see §3.

The above remarks are also valid for the last result mentioned in this section:

Theorem 6.4 [10]: Let M be a finite matroid, k natural number. Then there exists a finite matroid M' such that for every partition of the underlying set of M' into k classes there exists a restriction of M' which is isomorphic to M and which is a part of one of the classes of the partition.

(The proof is based on a suitable amalgamation of M with respect to a sparse high chromatic hypergraph).

Perhaps the only partition theorem proved so far concerning partitions of non-singletons in algebras is the case of finite abelian groups. In [4] and [15] all the finite abelian groups G with respect to which the category Ab (of all finite abelian groups) has the G-partition property are characterized. They are precisely the groups $\bigoplus_p (\mathbb{Z}_p \alpha(p))^{n(p)}$ where for every p, $\min\{\alpha(p), n(p)\} = 1$.

The proof of this theorem is related to the geometrical analogues of Ramsey's theorem: the affine Ramsey theorem, the vector space Ramsey theorem and the projective Ramsey theorem.

These theorems were proved in [5] by a generalization of the methods of [4]. (The methods used in the proofs of these theorems (a detailed study of the Hales-Jewett category and its variations, see [3],[5],[6],[7],[8],[12] and [15]) are at present the strongest methods in this area. However, as remarked in the introduction, this line of research is not covered here. This section in particular of the survey is essentially incomplete.

§7 Applications to partitions of integers

In this section we mention a few partition theorems for the sets of integers with forbidden configurations which are analogous to results for graphs and set systems mentioned in §2.

1) Van der Waerden's theorem:

Let X, Y be two sets of positive integers. The mapping $f: X \longrightarrow Y$ is said to be *sequential* if there exists $d > 0$ such that

$$d(x-y) = f(x)-f(y) \quad \text{for all} \quad x,y \in X.$$

In response to P. Erdős' question, J. Spencer and the authors independently proved the following:

Theorem 7.1: Let r be a positive integer. Then for every finite set X of integers not containing an arithmetical progression of length r there exists a set Y which also does not contain an arithmetical progression of length r. Moreover for every positive integer k and for every mapping

$c: Y \longrightarrow \{1,2,\ldots,k\}$ there exists a sequential mapping $f: X \longrightarrow Y$ such that $c \circ f$ is a constant mapping.

2) Schur's and Rado's Theorems:

We call a set of integers a Schur set if for every partition of A into finitely many parts $A = A_1 \cup A_2 \cup \ldots \cup A_k$ there exist numbers x,y and an index $i \leq k$ such that $\{x,y, x+y\} \subseteq A_i$.

Schur's theorem simply asserts that the set of all integers is a Schur set. We have investigated the question of how "locally sparse" the Schur set may be. We shall use the following notion of a local density:

Definition 7.2: Let A be a set of positive integers.

For every positive integer n put

$$\rho_n^A = \mathrm{Max}\{\rho(B);\ B \in [A]^n\}$$

where for every $B \subseteq A$ we put

$$\rho(B) = |\{x,y\} \subseteq B;\ x+y \in A\}|$$

We call the number ρ_n^A an *outer density*.
The following holds:

Theorem 7.3: Let A be a Schur set. Then $\rho_n^A \geq n-1$ for every n and there exists $n_o \in N$ such that $\rho_n^A \geq n$ for every $n \geq n_o$. Moreover for every n_o there exists a Schur set such that $\rho_n^A = n-1$ for every $n < n_o$. Analogously the *inner density* σ_n^A of a set A may be defined as follows:

$$\sigma_n^A = \mathrm{Max}\{\sigma(B);\ B \in [A]^n\}$$

where

$\sigma(B) = |\{x,y\}; \{x,y, x+y\} \subseteq B|$

The following holds:

Theorem 7.4: For every Schur set there exists n_o such that

$\sigma_n^A \geq [\frac{n-1}{2}]$ for $n \leq n_o$ and

$\sigma_n^A > [\frac{n-1}{2}]$ for $n > n_o$.

Moreover for every n_o there exists a Schur set A such that

$\sigma_n^A = [\frac{n-1}{2}]$ for $n \leq n_o$ and

$\sigma_n^A > [\frac{n-1}{2}]$ for $n > n_o$.

The analogous questions applied to the Rado's theorem seem to be much more difficult and we have no results corresponding to Theorems 7.3 and 7.4.

For $B = \{b_1, b_2, \ldots, b_t\}$ put $\sum B = \{\sum_{i\in\tau} b_i, \emptyset \neq \tau \subseteq \{1,2,\ldots t\}\}$ - we are able to prove the following only:

Theorem 7.5: Let r be a power of 2. Then there exists a set of integers A which has the following properties:

(i) $\sum B \nsubseteq A$ for arbitrary $B \subseteq A$, $|B| = r+1$

(ii) For every partition $A = A_1 \cup A_2 \cup \ldots \cup A_k$ into any finite numbers of parts there exists an index i and a set $B', |B'| = r$ such that $\sum B' \subseteq A_i$.

Finally we sketch one result which is related to Rado's theorem. First we state

Definition 7.6: Let A be a set of integers. Denote by $\lambda(A)$ the set of all integers t such that there are $y_1, y_2, \ldots, y_t$ with $\{y_1, y_2, \ldots, y_t, y_1+y_2+\ldots+y_t\} \subset A$.

Theorem 7.7: For every positive integer k, there exists a set A of positive integers which has the following property:

(1) For every partition $A = A_1 \cup A_2 \cup \ldots \cup A_\ell$ into a finite number of classes there exists $x_1, x_2, \ldots, x_k$ and an index i such that $\{x_1, x_2, \ldots, x_k, x_1+x_2+\ldots+x_k\} \subseteq A_i$

(2) $\lambda(A) = \{k+j(k-1);\ j = 0,1,2,\ldots\}$

It can be shown that this is the best possible result as $\lambda(A) \supseteq \{k+j(k-1);\ j = 0,1,\ldots\}$ for every set of integers A having the property (1).

All the above theorems were proved using convenient representations of integers by sets and applying a suitable theorem stated in §2 above.

§8 Selective theorems

There are various results which are considered to be of the Ramsey type but which are not covered by the above definitions of a particular partition property (see §3 above). We mention here one such example which we have studied. While in the above definition of the Ramsey-type theorem the number of colours is bounded we shall discuss here some results dealing with unrestricted partitions. These results are a natural generalization of one of the forms of Dirichlet's principle, which may be stated as follows:

For every positive integer n there exists a positive integer N such that for every set X with at least N elements the following holds: for every mapping $c: X \longrightarrow X$ (i.e. for every partition of a set X into any number of parts) there exists a subset $Y \subseteq X$, $|Y| = n$ such that c restricted to the set Y is either constant or a 1-1 mapping. (Of course it suffices to put $N = (n-1)^2+1$). The first result of this type was proved by Erdős and Rado. To formulate this we need the following:

Definition 8.1: Let $A \subset [Y]^k$. The mapping $\phi: A \longrightarrow [Y]^k$ is called *canonical* if there exists a total ordering $\leq$ of Y and a set $T \subseteq \{1,2,\ldots k\}$ such that

$$C(\{m_1,m_2,\ldots,m_k\}) = C(\{m_1',m_2',\ldots,m_k'\})$$

for $\{m_1,m_2,\ldots,m_k\}, \{m_1',m_2',\ldots,m_k'\} \in A$, $m_1 < m_2 <\ldots< m_k$, $m_1' < m_2' <\ldots< m_k'$ if and only if $m_i = m_i'$ for all $i \in T$.

Theorem 8.2: (Erdős-Rado canonization lemma). For every positive integer n,k there exists a positive integer N such that for every set X with at least N elements and every mapping $\phi: [X]^k \longrightarrow [X]^k$ there exists a set $Y \subset X$ with at least n elements such that the mapping ϕ restricted to the set $[Y]^k$ is canonical. The infinite form of this theorem is given now. For every positive integer k and for every mapping $\phi: [\omega]^k \longrightarrow [\omega]^k$ there exists an infinite set $Y \subset \omega$ such that the mapping ϕ restricted to the set $[Y]^k$ is canonical. (ω denotes the set of positive integers). It can easily be seen that putting $k = 1$ in the above theorem one gets the statement of Dirichlet's principle. First we discuss some results related to this statement.

Definition 8.3: Let (X,ϕ) be a hypergraph and $c: X \longrightarrow X$ be a mapping. We say that an induced subhypergraph (Y,τ) of (X,ϕ) is *c-selective* if c restricted to Y is canonical, which in this case means that it is either constant or 1-1 mapping.

Let (Y,τ) be a hypergraph. The hypergraph (X,ϕ) is said to be a *selective hypergraph for* (Y,σ) if for any mapping $c: X \longrightarrow X$, (X,τ) contains an induced subhypergraph which is c-selective and isomorphic to (Y,τ).

The k-hypergraph (X,ϕ) (i.e. hypergraph with all edges of cardinality k) is said to be *selective* if it is selective for (Y,τ) where $|Y| = k$ and $\tau = \{Y\}$.

Let (Y,τ) be a graph of chromatic number k such that $|Y| = n$. It can be shown that the minimum chromatic number of a graph (X,ϕ) which is selective for (Y,τ) is $(n-1)(k-1)+1$.

The statement of Dirichlet's principle clearly asserts that the hypergraph formed by all k point subsets of a set with cardinality $(k-1)^2+1$ is selective. The question arises what is the minimum number $s(k)$ of k-tuples a k-hypergraph must have to be selective. P. Erdős proved that $s(k) = (1+o(1))^k k^k$.

<u>Definition 8.4</u>: Let K be a class of hypergraphs. K is said to be *selective* if for every $(Y,\tau) \in K$ there exists $(X,\phi) \in K$ which is selective for (Y,τ).
The following can be proved:

<u>Theorem 8.5</u>: Let $k \geq 2$ be a positive integer. Let U be a finite set of 2-connected k-hypergraphs. Then the class of all hypergraphs which contain no $A \in U$ as an induced subhypergraph has the selective property. (A hypergraph (X,ϕ) is said to be *2-connected* if the hypergraph $(X-\{x\}, \{S; x \notin S \in \phi\})$ is connected for each $x \in X$). This theorem was recently used for a construction of non-zero-dimensional atoms in the lattice of countable uniform spaces having some additional properties (see [5]).

The following generalization of a Van der Waerden theorem which was proved by Erdős and Graham may also be considered as a theorem which is of the selective type.

Theorem 8.6: Let k be a positive integer. Then there exists a positive integer N such that for every mapping $c: \{1,2,\ldots N\} \longrightarrow \{1,2,\ldots,N\}$ there exists an arithmetical progression $A = \{a_o, a_o+d, \ldots, a_o+(k-1)d\} \subset \{1,2,\ldots,N\}$ such that the mapping c restricted to the set A is either 1-1 or constant. Let us remember that there are no analogous generalizations of the theorems of Hales-Jewett, Rado and even Schur. Now we state some results which may be considered as generalizations of the Erdős-Rado canonization lemma. First some definitions.

Definition 8.7: We say that the hypergraph H is *selective* for G with respect to F if for every mapping $c: \binom{H}{F} \longrightarrow \binom{H}{F}$ there exists a subhypergraph G' of H isomorphic to G such that the mapping c restricted to $\binom{G'}{F}$ is canonical.

Let K be a class of hypergraphs, $F \in K$. K is said to have the *F-selective property* if for every $G \in K$ there exists $H \in K$ which is selective for G with respect to F.

The hypergraph $G = (V,E)$ is said to be *irreducible* if $\bigcup(e\times e;\ e\in E) = V\times V$ (i.e. if E covers all pairs) (see §2).

Theorem 8.8: Let U be a finite set of irreducible k-hypergraphs. Then the class of all k-hypergraphs which contain no $A \in U$ as an induced subhypergraph has the F-selective property if and only if either $F \simeq (X,\emptyset)$ or $F \simeq (X,[X]^k)$ for some X.

At the end of this chapter we mention one result which is the unrestricted form of Theorem 4.7 and may be considered as a generalization of the Erdős-Rado canonization lemma (a joint result with P. Pudlák).

Let α be a countable ordinal. The notion of an α-uniform family was introduced in §4. The family is said to be uniform if it is α-uniform for some $\alpha < \omega_1$.

Definition 8.9: Let S be a set of finite subsets of ω. Put $S^* = \{t;\ \exists s \in S,\ t \subset s\}$. The mapping $\phi: S \longrightarrow \omega$ is said to be *canonical* if there exists a uniform family T, $T \subset S^*$, and $f: S \longrightarrow T$ such that

(i) $f(s) \subseteq s$

(ii) $\phi(s_1) = \phi(s_2)$ iff $f(s_1) = f(s_2)$

for every $s, s_1, s_2 \in S$.

Theorem 8.10: Let S be a uniform family. Then for every $\phi: S \longrightarrow \omega$ there exists an infinite $A \subset \omega$ such that the mapping ϕ restricted to $[A]^{<\omega} \cap S$ is canonical.

§9 Remarks on infinite sets and topological spaces

Not much is known about the infinite extensions of graph theoretical results discussed in this paper. In this chapter we mention some results concerning this topic. We also attempt to indicate how the transfinite extension of the Ramsey theorem may be applied to a solution of a problem in topology. Finally we mention some "Ramsey topological theorems".

As far as we know the paper of Erdős, Hajnal and Posa [4] contains almost all that is known about partitioning of edges of infinite graphs and induced embeddings. The following is proved in [4].

Theorem 9.1: For every countable graph $G = (V,F)$ there exists a graph $H = (V,E)$ (whose cardinality is not larger than the continuum) such that for every positive integer k and every mapping $c: E \longrightarrow \{1,2,\dots,k\}$ there exists an embedding $\phi: G \longrightarrow H$ (i.e. a mapping such that $\{x,y\} \in F$ iff $\{\phi(x),\phi(y)\} \in E$) such that $c \circ \phi$ is a constant mapping. Furthermore, it is also proved in [4] that if G is locally finite then H may be chosen countable and that this is not true for $G = K\omega,\omega$. A bit surprisingly the methods of [4] do not allow us to extend the above results to graphs with larger cardinalities or to k infinite and the following problems are open:

<u>Problem 9.2</u>: Let G be a graph of cardinality $> \omega$. Does there exist a graph $H = (V,E)$ with the following property: for every mapping $c: E \longrightarrow \{1,2\}$ there exists an embedding $\phi: G \longrightarrow H$ such that $c \circ \phi$ is a constant mapping?

<u>Problem 9.3</u>: Let G be a finite graph and $K \geq \omega$. Does there exist a graph $H = (V,E)$ with the following property: for every mapping $c: E \longrightarrow K$ there exists an embedding $\phi: G \longrightarrow H$ such that $c \circ \phi$ is a constant mapping?

It is not difficult to see that for G of a special form - G can be represented using one type (see [7]) - such an H exists. It can also be proved that if G is a cycle then H exists and moreover may be chosen so that it doesn't contain odd cycles of shorter length (this is stated for C_5 in [3]).

One of the most prominent problems connected with Problem 9.3 is due to Erdős and Hajnal.

<u>Problem 9.4</u>: Does there exist a graph $H = (V,E)$ which doesn't contain K_4 but for every $c: E \longrightarrow \omega$ there is an embedding $\phi: K_3 \longrightarrow H$ such that $c \circ \phi$ is a constant mapping? Similarly to the finite case, problems connected

with the partition of vertices are easier and the following Ramsey-type properties were considered.

Definition 9.5: Let $\mathcal{G}$ be a class of graphs. We say that $\mathcal{G}$ has a *vertex partition property* if for every $G \in \mathcal{G}$ and every cardinal number K there exists $H \in \mathcal{G}$, $H = (V,E)$, such that for every mapping $c: V \longrightarrow K$ there exists an embedding $\phi: G \longrightarrow H$ with $c \circ \phi$ a constant mapping.

We say that $\mathcal{G}$ has a *selective property* if for every $G \in \mathcal{G}$ there exists $H \in \mathcal{G}$, $H = (V,E)$, such that for every mapping $c: V \longrightarrow V$ there exists an embedding $f: G \longrightarrow H$ where $c \circ \phi$ is either constant or 1-1.

We say that $\mathcal{G}$ has an *ordering property* if for every $G \in \mathcal{G}$ and linear well-ordering $\leq$ of its vertices there exists $H \in \mathcal{G}$ such that for every linear well-ordering $\preccurlyeq$ of the vertices of H there exists an embedding $\phi: G \longrightarrow H$ which is monotone mapping with respect to $\leq$ and $\preccurlyeq$.

In most cases if $\mathcal{G}$ has a selective property then $\mathcal{G}$ has a vertex partition property, too. The following holds:

Theorem 9.6: The class of all graphs has the vertex partition, selective and ordering properties.

Let k be a positive integer. Then the class of all graphs which do not contain a complete graph with k vertices has the vertex partition property.

Ramsey's theorem has various transfinite extensions which are not discussed in this paper. These results found many applications in various fields of mathematics including topology. A list of topological applications of Ramsey-type theorems may be found in [6],[12]. Here we shall sketch one additional result which is related to unrestricted partitions and to uniform spaces where applications of this kind have not yet been so frequent (other examples are given in [10], [11]).

The open covering $\mathcal{U}$ of a metric space (X,ρ) is a system of open subsets of X such that for every $x \in X$ there is a $U \in \mathcal{U}$ such that $x \in U$. The covering $\mathcal{U}$ is called uniform if there exists an $\varepsilon > 0$ such that for every $x \in X$ there is a $U \in \mathcal{U}$ which contains the ε-ball $B_\varepsilon(x) = \{y;\rho(x,y) < \varepsilon\}$.

By the theorem of A.H. Stone [14] each open covering $\mathcal{U}$ of an arbitrary metric space (X,ρ) has an open locally finite refinement $\mathcal{V}$ (i.e. there is an open covering $\mathcal{V}$ with the two following properties: (i) for each $x \in X$ there is a neighbourhood of X which meets only finitely many members of $\mathcal{V}$, (ii) for every $V \in \mathcal{V}$ there is an $U \in \mathcal{U}$ so that $V \subset U$).

The question was raised in [5] whether in Stone's theorem one may replace the open coverings by uniform ones (i.e. if every uniform cover of a metric space has a locally finite uniform refinement). This question was answered in [9] and [13], where it was shown that the space $\ell_\infty(K)$ doesn't have this property. Using the transfinite extension of the Erdős-Rado canonical lemma [1] another example of such a space may be given. Indeed the following may be proved:

Theorem 9.7: For every cardinal number α there exists a metric space (x,ρ) such that for each 1-bounded uniform cover $\mathcal{V}$ of X (i.e. diam $V \subseteq 1$ for every $V \in \mathcal{V}$) there is an $x \in X$ contained in at least α numbers of $\mathcal{V}$.
We shall give a proof of Theorem 9.7 here as it is quite easy. We shall use the following lemma which follows from the results proved in [1]:

Lemma 9.8: For every cardinal number α and for every positive integer k there exists a cardinal number $\mathcal{B}_k$ such that the following holds: for every mapping $f: [\mathcal{B}_k]^k \longrightarrow [\mathcal{B}_k]^k$ there exists $Y \subset \mathcal{B}_k$ $|Y| = \alpha^+$ such that f restricted to $[Y]^k$ is canonical. (For the definition of the canonical mapping see 8.1).

Sketch of the proof of Theorem 9.7: Take $\mathcal{B} = \sup_k \mathcal{B}_k$ and $X = \bigcup_{k=1}^{\infty} [\mathcal{B}]^k$ and for $x \in [\mathcal{B}]^n$, $y \in [\mathcal{B}]^m$ define

$$\rho(x,y) = \begin{cases} \infty & \text{if } n \neq m \\ \dfrac{|x-y|+|y-x|}{n} & \text{if } n = m. \end{cases}$$

Take a 1 bounded uniform cover $\mathcal{U}$ of X. From the uniformity of $\mathcal{U}$ follows the existence of $\varepsilon > 0$ such that for every $x \in X$, $B_\varepsilon(x) \subset U$ for some $U \in \mathcal{U}$. Now take k so large that $\frac{1}{k} < \frac{\varepsilon}{2}$. Denote by X_k the subspace of X induced on a set $[\mathcal{B}]^k$. Choose $f: X_k \longrightarrow \mathcal{U}$ such that $B_\varepsilon(x) \subset f(x)$ for every $x \in X_k$. According to Lemma 9.8 there is a set Y of cardinality α^+ such that the mapping f restricted to the set $[Y]^k$ is canonical. The corresponding T (from the definition 8.1 of canonical mapping) must be nonempty as it follows from $T = \emptyset$ that there is $U \in \mathcal{U}$ such that $f(y) = U$ for every $y \in [Y]^k$ and this contradicts the fact that the diameter of the space induced on a set $[Y]^k$ is 2 while diam $U < 1$.

Choose $Z = \{\{a_1, a_2, \ldots, a_{\text{Min } T} + \gamma, \ldots, a_k\} \subset [Y]^k$ where

$$a_1 < a_2 < \ldots < a_{\text{Min } T} < a_{\text{Min } T} + \alpha < a_{\text{Min } T+1} < a_k.$$

Fix $z_0 \in Z$; we have $z_0 \in f(z)$ for every $z \in Z$. As $|Z| = \alpha$ the theorem is proved. Let us close this part with the following remark. We may also ask the direct question, has the category of all topological spaces a point partition property? i.e. whether for every topological space X and cardinal γ there exists a topological space Y such that for every mapping $c: Y \longrightarrow \gamma$ there exists a homeomorphism $\phi: Y \longrightarrow Y$ where $\phi \circ c$ is a constant mapping. The following holds:

Theorem 9.9 [8]: The class of all topological spaces (resp. T_o, T_1 spaces) has the point partition, selective and ordering properties. (Selective and ordering properties are defined similarly as for graphs).
The above theorem cannot be extended to T_2 spaces as we have:

Theorem 9.10 [15]: Every T_2 space can be partitioned into two parts such that neither of those parts contains the space of reals as a subspace.

§10 Concluding remarks

Finally we shall sketch a few related results which are not Ramsey-type statements but which are closely connected to some of the results considered in the previous sections, particularly to the following strengthening of Ramsey's theorem which was one of the first steps in the direction principally studied by the authors [1,4,9]:

Theorem 10.1: For every finite graph G there exists a graph $F = (V(F),E(F))$ such that for every partition of edges of the graph F into two parts $E(F) = E_1 \cup E_2$ there exists an induced subgraph G' of F, G' isomorphic to G, such that all edges of G' belong either to E_1 or to E_2.

Notice that all theorems given in §2 extend Theorem 10.1. In [6] the authors gave a simple proof of Theorem 10.1 using the fact that every finite graph is an induced subgraph of a direct product of complete graphs. Here we state two other theorems which imply the universality of a convenient class of graphs and thus provide independent proofs of Theorem 10.1.

In [3] Erdős and Hajnal stated the following:

Theorem 10.2: Let $e(n,k) = \frac{1}{2}\left(\frac{1}{2k+1}\log n\right)^{\frac{1}{2}}$ and $p(n,k) = 2^{e(n,k)}$. There is a function $n(k)$ such that whenever H is a graph on $n > n(k)$ vertices and neither H nor its complement contain a complete graph on $p(n,k)$ vertices then every graph of order k is an induced subgraph of H.

We sketch how Theorem 10.1 follows from Theorem 10.2. For convenience in the following we identify H with its edge set.

Let G be a given graph on k vertices. Consider a graph F on n vertices such that neither F nor its complement contain a complete graph on $[2 \log_2 n]$ vertices (the existence of such a graph was proved by Erdős in 1959 by probabilistic means [2]) and assume that n is large enough to justify all computations below.
Let $F = F_1 \cup F_2$ be a partition of edges of a graph F.
Suppose that the complement of the graph F_1 contains a clique on at least $p(n,k)$ vertices. (If it were not so G would be an induced subgraph of F_1 as F_1 does not contain a clique of the size $p(n,k)$).
Thus there exists an induced subgraph H of F_2 on $m \geq p(n,k)$ vertices such that neither H nor its complement contains a complete subgraph on $2 \log_2 n$ vertices.
As $p(m,k) > 2 \log n$ for n sufficiently large, the proof is finished.

Another statement which under certain conditions on a graph implies universality of a graph may be proved using a lemma of Szemerédi. We shall state this lemma here as it is an

important and interesting statement which has various applications. First we need some definitions:
Let A,B be nonempty disjoint subsets of the vertex set of a graph $G = (V,E)$.

Define $d(A,B) = \frac{|E(A,B)|}{|A||B|}$, where $E(A,B)$ denotes the set of all edges of G with one endpoint in A and another in B.

The pair (A,B) is called *ε-regular* if $X \subseteq A$, $Y \subseteq B$, $|X| \geq \varepsilon|A|$, $|Y| \geq \varepsilon|B|$ imply $|d(X,Y) - d(A,B)| < \varepsilon$.

The partition $V = C_o \cup C_1 \cup \ldots \cup C_k$ of the vertex set of C is said to be *ε-regular* if

(i) $|C_o| \leq \varepsilon|V|$

(ii) $|C_1| = |C_2| = \ldots = |C_k|$

(iii) at least $(1-\varepsilon)\binom{k}{2}$ pairs C_i, C_j, $1 \leq i < j \leq k$ are ε-regular.

Theorem 10.3: (Szemerédi [10]). For every positive real ε and for every positive integer m there are positive integers $N = N(\varepsilon,m)$ and $M = M(\varepsilon,m)$ with the following property: for every graph $G = (V,E)$ with at least N vertices there is an ε-regular partition of G into $k+1$ classes such that $m \leq k \leq M$. Before we present another statement which may be considered as an extension of Theorem 10.1 we shall need one more definition:

Let γ,σ be given reals, $0 < \gamma < 1$, $0 < \delta < \frac{1}{2}$. We say that the graph $H = (W,F)$ has the *property* (γ,δ) if the number of edges of each subgraph induced on a set S with cardinality $\geq \gamma|W|$ is bigger than $\delta\binom{|S|}{2}$ and smaller than $(1-\delta)\binom{|S|}{2}$.

Theorem 10.4: For every positive integer ℓ and for every real δ, $0 < \delta < \frac{1}{2}$, there exist $\gamma = \gamma(\ell,\delta) > 0$ and $N_o = N_o(\ell,\delta)$ such that every graph H with at least N_o vertices which has the property (γ,δ) contains all graphs with ℓ vertices as induced subgraphs.

We shall omit the proof of Theorem 10.4 but do give another simple proof of Theorem 10.1 now using Theorem 10.4. In fact, we shall prove the following slightly weaker statement:

Theorem 10.1': For every finite graph G there exists a graph $F' = (V(F'),E(F'))$ such that for every partition of edges of the graph F' into two parts $E(F') = E_1 \cup E_2$ there exists $i \in \{1,2\}$ such that G is an induced subgraph of $(V(F'),E_i)$.

The fact that F' has the property stated above is denoted (for this proof only) by $F' \rightsquigarrow G$.

Theorem 10.1' is in fact only seemingly weaker as one can easily prove the validity of Theorem 10.1 using Theorem 10.1'. Take L such that $L \longrightarrow (\ell)^2_2$ holds (i.e. $L \geq r(\ell,2,2)$, see §1). For a graph G take the graph F' from Theorem 10.1'. Put $F' = F_o$ and define the graphs $F_1, F_2, \ldots, F_{\binom{L}{2}}$ as follows: let F_i be given, take F_{i+1} as a graph for which $F_{i+1} \rightsquigarrow F_i$ holds. One can easily check now that the product $K_L \times F_{\binom{L}{2}}$ has the property of F in Theorem 10.1. (The product of two graphs (V_1,E_1), (V_2,E_2) is here understood to be the graph with vertex set $V_1 \times V_2$ and with two vertices $\langle v_1,v_2\rangle$, $\langle w_1,w_2\rangle$ joined by an edge if both $\{v_1,w_1\} \in E_1$ and $\{v_2,w_2\} \in E_2$.) It remains to prove the implication "Theorem 10.4 $\Longrightarrow$ Theorem 10.1".

Let G be a given graph with ℓ vertices. Put $N_o = N_o(\ell, \frac{1}{20})$

$$\gamma_1 = \mathrm{Min}\{\gamma(\ell, \tfrac{1}{20}), \tfrac{1}{10}\}$$

Take ε so small that $r([1/\varepsilon]) > (\frac{1}{\gamma_1})^2$ where $r(n)$ is the largest t such that any 2-colouring of $[n]^2$ yields a monochromatic $[S]^2$, $|S| \geq t$.

Put $m \geq [1/\varepsilon]$.

Consider now the graph F' on N vertices

$(N > \mathrm{Max}\{N(\varepsilon,m); N_o.M(\varepsilon,m) \frac{\gamma_1^2}{1-\varepsilon}\})$ such that for every two disjoint sets A,B of cardinality at least $\frac{\varepsilon N(1-\varepsilon)}{M(\varepsilon,m)}$, $d(A,B) \in (\frac{2}{3} - \frac{\varepsilon}{2}, \frac{2}{3} + \frac{\varepsilon}{2})$.

The existence of such a graph may be shown by a simple counting argument. Let $F_1 \cup F_2$ be a partition of edges of the graph F'. Consider an ε-regular partition $V = C_o \cup C_1 \cup \ldots \cup C_k$ of the vertices of the graph F_1 where $M(\varepsilon,m) \geq k \geq m$. (Such a partition exists according to Theorem 10.3.) As $(1-\varepsilon)\binom{k}{2}$ of pairs $(C_j,C_{j'})$ $1 \leq j < j' \leq k$ are ε-regular it follows from Turán's theorem [11] that there are $[1/\varepsilon]$ classes, $C_{i_1}, C_{i_2} \ldots C_{i_{[\frac{1}{\varepsilon}]}}$, such that all pairs $(C_{i_\alpha}, C_{i_\beta})$ are ε-regular $\alpha \neq \beta$, $\alpha,\beta \in \{1,2,\ldots,r[1/\varepsilon]\}$. It follows from Ramsey's theorem that there exist sets $A_1 = C_{i_{\alpha_1}}$, $A_2 = C_{i_{\alpha_2}}, \ldots, A_{r[1/\varepsilon]} = C_{i_{\alpha_{r[1/\varepsilon]}}}$ such that either $d_{F_1}(A_j,A_{j'}) > \frac{1}{3}$ for all $j \neq j'$, $j,j' \in \{1,2,\ldots,r[1/\varepsilon]\}$ or $d_{F_1}(A_j,A_{j'}) \leq \frac{1}{3}$ for all $j \neq j'$ $j,j' \in \{1,2,\ldots,r[1/\varepsilon]\}$. Suppose that $d_{F_1}(A_j,A_{j'}) = \rho$ then

$$d_{F_2}(A_j, A_{j'}) \in (\frac{2}{3} - \rho - \frac{\varepsilon}{2}, \frac{2}{3} - \rho + \frac{\varepsilon}{2})$$

and moreover

$$d_{F_2}(X_j, X_j') \in (\frac{2}{3} - \rho - \frac{3\varepsilon}{2}, \frac{2}{3} - \rho + \frac{3\varepsilon}{2})$$

for every

$$|X_j| \geq \varepsilon |A_j|, \quad |X_j'| \geq \varepsilon |A_j'| \tag{1}$$

Thus we can suppose that there is an $i \in \{1,2\}$ such that

$$\rho_{jj'} - \frac{3}{2}\varepsilon < d_{F_i}(X_j, X_{j'}) < \rho_{jj'} + \frac{3}{2}\varepsilon \tag{2}$$

and

$$\frac{2}{3} \geq \rho_{jj'} \geq \frac{1}{3}$$

for all $j, j' \in \{1, 2, \ldots, r[\frac{1}{\varepsilon}]\}$ and all $X_j \subset A_j$, $X_{j'} \subset A_{j'}$ satisfying (1).

Consider now the subgraph T of a graph F_i induced on a set $A = \bigcup_{i=1}^{r[1/\varepsilon]} A_i$.

A simple computation shows that for any set $B \subset A$ such that $|B| > \gamma_1 |A|$, the number of vertices of B which belong to those A_i for which $|A_i \cap B| > \varepsilon |A|$ holds is larger than $\frac{9}{10}|B|$ and moreover that there are at least 9 A_i's such that $|A_i \cap B| > \varepsilon |A_i|$.

From this and (2) it follows that T has the property $(\gamma_1, \frac{1}{20})$ and as it has at least $(1-\varepsilon)\frac{r[1/\varepsilon]}{M(\varepsilon,m)} N > N_o$ vertices it contains G as an induced subgraph.

We shall close this paper with the following problem.

Problem 10.5: For a given graph $G = (V,E)$ determine the smallest number $r(G)$ of vertices of a graph F for which the statement of Theorem 10.1 holds.

None of the above proofs of Theorem 10.1 yields a reasonable bound for $r(G)$. It is not even known whether $r(G) > r(|V|,2,2)$ for a graph G.

The purpose of this paper was to demonstrate some mathematical connections of partition (Ramsey) theory. The paper covers a relatively broad range of topics (although not as broad as were the interests of F.P. Ramsey himself). D. Gale recently [5] drew attention to Ramsey's paper [8] which is one of the fundamental papers in mathematical economics.

References

§1

[1] N.G. de Bruijn, P. Erdős; Indag. Math. 13 (1951) 369-373

[2] S. Burr; in Graphs and Combinatorics (eds. R.A. Bari and F. Harary). Lecture Notes in Maths. No.406 Springer (1973)

[3] S. Burr; Ann. of N.Y. Acad. Sci. (to appear)

[4] W. Deuber; Math. Z. 133 (1973) 109-123

[5] R. Graham, B. Rothschild; in Combinatorics (M. Hall Jr. and J.H. van Lint, eds.) Reidel (1975)

[6] R. Graham, K. Leeb, B. Rothschild; Adv. in Maths. 8 (1972) 417-433

[7] J. Nesetril, V. Rődl in Coll. Math. Soc. János Bolyai 18, 759-792 Noth-Holland (1978)

[8] R. Rado; Math. Z. 36 (1973) 424-480

[9] F.P. Ramsey; Proc. Lond. Math. Soc. 30 (1930) 264-286

[10] I. Schur; J.Deutsch. Math. Verein 25 (1916) 114

[11] E. Szemerédi; Acta Arithmetica 27 (1975) 199-245

[12] B.L. Van der Waerden; Nieuw. Arch. Wisk. 15 (1928) 212-216

[13] P. Erdős, R. Rado; Proc. Lond. Math. Soc 3 (1951) 417-439

[14] P. Erdős, A. Hajnal, R. Rado; Acta Math. Acad. Sci. Hung. 16 (1965) 93-196

§2

[1] F.G. Abramson, L.A. Harrington; J. Symbolic Logic 43 (1978) 572-600

[2] C. Berge; Graphs and Hypergraphs, North-Holland (1970)

[3] S.A. Burr, P. Erdős, L. Lovász; Ars. Combinatoria 1 (1976) 167-190

[4] P. Erdős in Recent Advances in Graph Theory pp.183-192 (M. Fiedler, ed.) Academia, Prague (1975)

[5] J. Nešetril, A. Pultr; Discrete Math. 22 (1978) 287-300

[6] J. Nešetril, V. Rödl in Recent Advances in Graph Theory pp. 405-412 (M. Fiedler, ed.) Academia, Prague

[7] J. Nešetril, V. Rödl in Coll. Math. Soc. János Bolyai 10, 1127-1132 North-Holland (1975)

[8] J. Nešetril, V. Rödl; J. Comb. Th. B 20 (1976) 243-249

[9] J. Nešetril, V. Rödl; J. Comb. Th. A 22 (1977) 289-312

[10] J. Nešetril, V. Rödl in Coll. Math. Soc. János Bolyai 18, 759-792 North-Holland (1978)

[11] J. Nešetril, V. Rödl; Proc. Am. Math. Soc. 72 (1978) 417-421

[12] J. Nešetril, V. Rödl; Math. Slovaca (to appear)

[13] B.L. Rothschild (preprint)

[14] J. Spencer; J. Comb. Th. A 19 (1975) 278-286

§3

[0] W. Deuber; Math. Z. 133 (1973) 109-123

[1] R. Graham, B. Rothschild; Trans. Am. Math. Soc. 159 (1971) 257-292

[2] R. Graham, K. Leeb, B. Rothschild; Adv. in Maths 8 (1972) 417-433

[3] R. Graham, B. Rothschild in Combinatorics (M. Hall Jr. and J.H. van Lint, eds.) pp.261-277 Reidel (1975)

[4] A. Hales, R.I. Jewett; Trans. Am. Math. Soc. 106 (1963) 222-229

[5] K. Leeb; Vorlesungen über Pascaltheorie Erlangen (1973)

[6] K. Leeb in Coll. Math. Soc. Janós Bolyai 10 North-Holland (1975)

[7] J. Nešetril, V. Rödl in Recent Advances in Graph Theory pp.405-412 (M. Fiedler, ed.) Academia, Prague (1975)

[8] - , - Ibid. pp.413-423

[9] - , - Bull. Am. Math. Soc. 83 (1977) 127-128

[10] J. Nešetřil, V.Rödl; J. Comb. Th. A 22 (1977) 289-312

[11] - , - Proc. Am. Math. Soc. 72 (1978) 417-421

[12] - , - Math. Slovaca (to appear)

[13] - , - in Coll. Math. Soc. János Bolyai 18, 754-792 North-Holland (1978)

[14] R. Rado; Math. Z. 36 (1933) 424-480

[15] F.P. Ramsey; Proc. Lond. Math. Soc. 30 (1930) 264-286

[16] J. Spencer (to appear)

[17] B. Voigt (to appear)

§4

[1] F.G. Abramson, L.A. Harrington; J. Symbolic Logic 43 (1938) 572-600

[2] J. Edmonds; Can. J. Math. 17 (1965) 449-467

[3] P. Erdös, G. Mills (preprint)

[4] P. Erdös, R. Rado; Proc. Lond. Math. Soc. 3 (1951) 417-439

[5] C.G. Jockusch; J. Symbolic Logic 37 (1972) 268-280

[6] J. Ketonen (preprint)

[7] C. St.J.A. Nash-Williams; Proc. Camb. Phil. Soc. 61 (1965) 33-39

[8] J. Nešetřil, V. Rödl; Bull. Am. Math. Soc. 83 (1973) 127-128

[9] - , - J. Comb. Th. A 22 (1977) 289-312

[10] J. Paris (preprint)

[11] J. Paris, L. Kirby in Proc. of the Bierntowice Conference Springer (1976)

[12] J. Paris, L. Harrington in Handbook of Math. Logic pp. 1133-1142 North-Holland (1977)

[13] P. Pudlák (to appear)

[14] J.R. Schoenfield; Mathematical Logic, Addison-Wesley (1967)

[15] J. Silver; (preprint)

[16] R. Solovay; (preprint)

§5

[1] J. Baumgartner; J. Comb. Th. A 17 (1974) 384-386

[2] J. Baumgartner, A. Taylor; Trans. Am. Math. Soc. 241 (1978) 283-311

[3] N.G. de Bruijn, P. Erdős, Indag. Math. 13 (1951) 369-373

[4] W.W. Comfort; Bull. Am. Math. Soc. 83 (1977) 449-472

[5] W.W. Comfort, S. Negrepontis; Theory of Ultrafilters, Springer (1974)

[6] F. Galvin (private communication)

[7] R.L. Graham, B.L. Rothschild; Trans. Am. Math. Soc. 159 (1971) 257-292

[8] N. Hindman; J. Comb. Th. A 17 (1974) 1-11

[9] - ; Notices Am. Math. Soc. 25 (1978) # 759-A1

[10] K. Kunen; Proc. Camb. Phil. Soc.

[11] K. Leeb; Vorlesungen über Pascaltheorie, Erlangen (1973)

[12] J. Nesetril; Comm. Math. Univ. Carol. 18 (1977) 675-683

[13] J. Nesetril, V. Rödl; J. Comb. Th. A 22 (1977) 289-312

§6

[1] B. Bollobás; Alg. Universalis 7 (1977) 313-314

[2] W. Deuber; J. Comb. Th. A 19 (1975) 95-108

[3] W. Deuber, B. Rothschild in Coll. Math. Soc. János Bolyai 18, North-Holland (1978)

[4] R. Graham, B. Rothschild; Trans. Am. Math. Soc. 159 (1971) 251-292

[5] R. Graham, K. Leeb, B. Rothschild; Adv. in Maths. 8 (1972) 417-433

[6] A. Hales, R.I. Jewett; Trans. Am. Math. Soc. 106 (1963) 222-229

[7] K. Leeb; Vorlesungen über Pascaltheorie, Erlangen (1973)

[8] J. Nešetril, V. Rödl; submitted to Alg. Universalis

[9] J. Nešetril, S. Poljak, D. Tursik; submitted to J. Comb. Th.

[10] P. Pudlák, J. Tůma; Alg. Universalis (to appear)

[11] J. Spencer; (to appear)

[12] I. Schur; J. Deutsch Math. Verein 25 (1916) 114

[13] R. Rado; Math. Z. 36 (1933) 424-480

[14] B. Voigt (to appear)

[15] J. Nešetril, V. Rödl; Proc. Am. Math. Soc. 72 (1978) 417-421

§7

[1] J. Nešetril, V. Rödl; Comm. Math. Univ. Carol. 17 (1976) 675-682

[2] - , - (to appear)

[3] J. Spencer; J. Comb. Th. A 19 (1975) 278-286

§8

[1] P. Erdős (private communication)

[2] P. Erdős, R. Graham (private communication)

[3] P. Erdős, R. Rado; Proc. Lond. Math. Soc. 3 (1978) 181-189

[4] J. Nešetril, V. Rödl; Ann. of Discrete Maths. 3 (1978) 181-189

[5] J. Pelant, J. Reiteman, V. Rödl, P. Simon (to appear)

§9

[1] J. Baumgartner; J. Symbolic Logic 40 (1975) 541-554

[2] P. Erdős in Recent Advances in Graph Theory (M. Fiedler,, ed.) pp.183-192, Academia Prague (1975)

[3] P. Erdős, A. Hajnal; Ann. of N.Y. Acad. Sci. 175 (1970) 115-124

[4] P. Erdős, A. Hajnal, L. Posa in Coll. Math. Soc. Janós Bolyai 10, 1127-1132, North-Holland

[5] J.R. Isbell; Uniform Spaces, Am. Math. Soc. (1964)

[6] I. Juhász; Cardinal functions in topology, Math. Centre Tracts, Amsterdam (1974)

[7] J. Nešetril, V. Rödl in Recent Advances in Graph Theory (M. Fiedler, ed.) pp.405-412 Academia, Prague (1975)

[8] - , - in Proc. of the 4th Prague Toposymposium, pp.333-337 Prague (1977)

[9] J. Pelant in Seminar on Uniform Spaces 1973/74 pp.149-158 (Z. Frolik, ed.) Prague 1974

[10] - in Seminar on Uniform Spaces 1975/76, pp.145-150 (Z. Frolik, ed.) Prague 1976

[11] J. Pelant, V. Rödl (to appear)

[12] M.E. Rudin; Lectures on Set Theoretic Topology, Am. Math. Soc. (1975)

[13] E.V. Shepin; Dokl. Acad. Nauk. U.S.S.R. 222 (1975) 541-543

[14] A.H. Stone; Bull. Amer. Math. Soc. 54 (1948) 977-982

[15] W. Weiss, Lecture at the Hungarian Topological Colloquium, Budapest, 1978

§10

[1] W. Deuber in Coll. Math. Soc. Janós Bolyai 10, North-Holland (1975)

[2] P. Erdős, Canad. J. Math. 11 (1959) 34-38

[3] P. Erdős, A. Hajnal in Beiträge sur Graphen Theorie pp.80-96, Technische Hochschule Ilmenau (1977)

[4] P. Erdős, A. Hajnal, L. Posa in Coll. Math. Soc. Janós Bolyai 10, 1122-1132, North-Holland

[5] D. Gale, Bull. Am. Math. Soc. 84 (1978) 669

[6] J. Nesetril, V. Rödl; Disc. Math. 23 (1978) 49-55

[7] F.P. Ramsey; Proc. Lond. Math. Soc. 30 (1930) 264-286

[8] - , Economic Journal 38 (1928) 543-559

[9] V. Rödl in Graphs, Hypergraphs and Block Systems pp.211-220, Zielona Gora (1976)

[10] E. Szemerédi in Coll. Internationaux C.N.R.S. 260 399-402 Paris (1978)

[11] P. Turán; Mat. Fiz. Zapok 48 (1941) 436-452

6 · Strongly regular graphs

J.J. Seidel

1. INTRODUCTION

The present paper is meant to give a selfcontained introduction to the theory (rather than to the construction) of strongly regular graphs, with an emphasis on the configurational aspects and the Krein parameters.

The definition in graph-theoretical terms in §2 is translated into the language of matrices and eigenvalues in §3, leading to a short arithmetic of the parameters in §4. In §5 the adjacency algebra relates the combinatorial and the algebraic idempotents, leading to spherical 2-distance sets and the Krein inequalities. The absolute bound of §6 provides inequalities of a different type. Necessary and sufficient conditions for equality in either bound are considered in §7, as well as their graph-theoretic significance. Finally, the notion of switching is explained in §8, and illustrated by two explicit examples in §9.

Strongly regular graphs have been introduced by Bose [1], related to algebra by Bose and Mesner [2], to eigenvalues by Hoffman [11], and to permutation groups by D.G. Higman [10]. Scott [17] introduced the Krein parameters. Margaret Smith [20] considered extremal rank 3 graphs, providing a motivation for the recent work by Cameron et al. [7], whose pattern was followed in the present paper. For a survey of construction methods for strongly regular graphs the reader is referred to Hubaut [12], and for relations with other combinatorial objects to Cameron - van Lint [5], Cameron [6] and Goethals-Seidel [8],[9].

2. DEFINITIONS

We shall deal with ordinary graphs (undirected, without loops and multiple edges) having a finite number n of vertices. For any pair of vertices a *between-vertex* is a vertex adjacent to both. Let k, λ, μ be any integers such that

(i)' each vertex is adjacent to at most k other vertices,

(ii)' each adjacent pair of vertices has at least λ between-vertices,

(iii)' each non-adjacent pair of vertices has at least μ between-vertices.

Then we claim that

$$(n-k-1)\mu \leq k(k-1-\lambda) \quad .$$

Indeed, select any vertex p and denote by $\Gamma(p)$ and $\Delta(p)$ the sets of vertices adjacent and non-adjacent to p, respectively. Now p and any $y \in \Delta(p)$ have at least μ between-vertices in $\Gamma(p)$, hence there are at least $(n-k-1)\mu$ edges between $\Gamma(p)$ and $\Delta(p)$. Furthermore, p and any $x \in \Gamma(p)$ have at least λ between-vertices in $\Gamma(p)$, hence there are at most $k(k-1-\lambda)$ edges between $\Gamma(p)$ and $\Delta(p)$. This proves the formula above. In addition, this shows that

$$(n-k-1)\mu = k(k-1-\lambda)$$

if and only if the following hold:

(i) each vertex is adjacent to k other vertices,

(ii) each adjacent pair of vertices has λ between-vertices,

(iii) each non-adjacent pair of vertices has μ between-vertices.

A graph is called *strongly regular* whenever there exist integers k, λ, μ satisfying (i), (ii), (iii). Clearly, if a graph is strongly regular then so is its complement. For reasons of presentation we shall exclude graphs consisting of a disjoint union of complete subgraphs of equal size, and their complements. Thus, our strongly regular graphs and their complements are connected (have $\mu \neq 0$).

<u>Example</u> The pentagon graph is strongly regular with

$$n = 5,\ k = 2,\ \lambda = 0,\ \mu = 1 \quad .$$

<u>Example</u> The Petersen graph is strongly regular with

$$n = 10,\ k = 3,\ \lambda = 0,\ \mu = 1 \quad .$$

<u>Example</u> Let V be a set of $v \geq 5$ elements. The triangular graph $T(v)$ is the graph whose vertices are the unordered pairs from V , two vertices being adjacent whenever the corresponding pairs have an element in common. $T(v)$ is strongly regular with

$$n = \tfrac{1}{2}v(v-1),\ k = 2(v-2),\ \lambda = v-2,\ \mu = 4 \quad .$$

Strongly regular graphs arise for instance in connection with rank 3 permutation groups, cf. [10]. Let G denote a permutation group on the set Ω . On $\Omega \times \Omega$ the relation

$$\forall_{\alpha,\beta,\gamma,\delta\in\Omega}((\alpha,\beta) \sim (\gamma,\delta) \ :\Longleftrightarrow\ \exists_{g\in G}(\gamma = g(\alpha),\ \delta = g(\beta)))$$

is an equivalence relation. If Δ is an equivalence class, then so is

$$\Delta^{\cup} := \{(\beta,\alpha) \mid (\alpha,\beta) \in \Delta\} \quad .$$

G is transitive on Ω iff the diagonal I of $\Omega \times \Omega$ is an equivalence class. G is 2-transitive on Ω iff I and its complement in $\Omega \times \Omega$ are the only equivalence classes. The 'next' case is that of 3 classes:

$$\Omega \times \Omega = I \cup \Gamma \cup \Delta \ .$$

Then either $\Gamma^{\cup} = \Gamma$, $\Delta^{\cup} = \Delta$ or $\Gamma^{\cup} = \Delta$, $\Delta^{\cup} = \Gamma$, and these cases occur according as the order of G is even or odd. In the even case the sets $\Gamma = \Gamma^{\cup}$ and $\Delta = \Delta^{\cup}$ define the edges of two complementary (undirected) graphs on the vertex set Ω. These *rank 3 graphs* are strongly regular since G is transitive on vertices (k is constant), on edges (λ is constant), and on non-edges (μ is constant). We shall not deal with the odd case.

Example The triangular graph $T(v)$ arises from the action of the symmetric group on v symbols on the set of unordered pairs of symbols. The Petersen graph is the complement of $T(5)$.

3. ADJACENCY MATRICES

Let A and B denote the $(1,0)$ adjacency matrix of a graph and of its complement, then

$$A + B + I = J \ ,$$

where I and J denote the unit and the all-one matrix of size n, respectively. The definition of a strongly regular graph translates into

$$A^2 = kI + \lambda A + \mu B \ .$$

Indeed, the diagonal entries of A^2 equal k, and the

off-diagonal entries equal λ and μ according as the corresponding two vertices are adjacent and non-adjacent, respectively. The matrix equation above yields

$$A^2 + (\mu-\lambda)A - (k-\mu)I = \mu J \quad , \qquad AJ = kJ \quad ,$$

and, by factorization of the quadratic form in A ,

$$(A-rI)(A-sI) = \frac{(k-r)(k-s)}{n} J \quad .$$

Hence, apart from the eigenvalue k (with eigenvector the all-one vector j) the adjacency matrix A has the eigenvalues r and s . Due to our restriction to $\mu \neq 0$ for the graph and for its complement, the eigenvalues k,r,s are distinct. Conversely, if the adjacency matrix A of any graph has 3 distinct eigenvalues, one of which with multiplicity one and eigenvector j , then the graph and its complement are strongly regular with $\mu \neq 0$. Thus, we have equivalent definitions for strongly regular graphs in terms of the adjacency matrix, and in terms of the eigenvalues of the adjacency matrix.

Example The adjacency matrix A of the triangular graph $T(v)$ satisfies

$$(A+2I)(A-(v-4)I) = 4J \quad , \qquad AJ = 2(v-2)J \quad .$$

Certain combinatorial objects give rise to strongly regular graphs. A *quasi-symmetric* block design, cf. [8], [5], is a block design $2\text{-}(V,K,\Lambda)$ with the property that the blocks have two intersection numbers, that is, there exist integers x and y such that any two distinct blocks intersect either in x or in y blocks. Let N denote the point-block incidence matrix, then

$$NN^t = (R-\Lambda)I + \Lambda J \quad , \qquad NJ = RJ \quad , \qquad JN = KJ \quad ,$$

$$N^tN = KI + xA + yB \quad , \qquad I + A + B = J \quad ,$$

where A and B are the characteristic matrices, of size $V(V-1)/K(K-1)$, of the relations on the set of blocks defined by intersection in x and in y points, respectively. Since NN^t and N^tN (possibly matrices of different size) have the same eigenvalues $\neq 0$ with the same multiplicities, it follows that A and B are the adjacency matrices of a pair of complementary strongly regular graphs.

Example The triples of a Steiner triple system on v symbols, $v > 9$, form a strongly regular graph with $x = 1$, $y = 0$ and

$$n = \frac{1}{6}v(v-1) \ , \ k = \frac{3}{2}(v-3) \ , \ r = \frac{1}{2}(v-9) \ , \ s = -3 \quad .$$

A partial geometry, cf. [1], with K points per line, R lines per point, and T points per line connected with any point not on that line, gives rise to two strongly regular graphs. Their adjacency matrices P (of the point graph) and L (of the line graph) are related via the point-line incidence matrix N by

$$NN^t = P + RI \quad , \qquad N^tN = L + KI \quad .$$

For $K = T = 3$ we are back in the case of Steiner triple systems. Also other combinatorial objects give rise to strongly regular graphs, such as sets of mutual orthogonal Latin squares, and regular symmetric Hadamard matrices with a constant diagonal, cf. [5], [8].

4. ARITHMETIC

Let the adjacency matrix A of a strongly regular graph have the eigenvalues k,r,s with the multiplicities $1,f,g$, respectively, and let $r > s$. The adjacency matrix B of the complementary graph has the eigenvalues $\ell := n-1-k, -r-1, -s-1$ with the multiplicities $1,f,g$. From trace $A = 0$, trace $A^2 = nk$, and from the definitions, the following relations between the parameters are derived:

$$k + fr + gs = 0 , \quad k^2 + fr^2 + gs^2 = nk ,$$

$$(n-k-1)\mu = k(k-1-\lambda) , \quad 1 + f + g = n , \quad \lambda - \mu = r + s ,$$

$$\mu = k + rs = (k-r)(k-s)/n , \quad \ell(k+rs) = -k(r+1)(s+1) ,$$

$$f = \frac{-k(s+1)(k-s)}{(k+rs)(r-s)} , \quad g = \frac{k(r+1)(k-r)}{(k+rs)(r-s)} .$$

These relations imply restrictions for the possible values of the parameters. For instance, $k + fr + gs = 0$ implies

$$2k + (n-1)(r+s) + (f-g)(r-s) = 0 ,$$

whence $r,s \in \mathbb{Z}$ for $f \neq g$, and $r + s = -1$ for $f = g$. Moreover, divisibility conditions are obtained from the fact that n,k,λ,μ,f,g are nonnegative integers. For instance, in the special case $\mu = \lambda + 1$, $r \in \mathbb{Z}$, elimination of s,n,k from the equations above yields

$$(2r+1)^4 - 2(2r+1)^2 - 16(f-g)(2r+1)\mu - 16\mu^2 + 1 = 0 ,$$

whence $2r + 1$ divides $16\mu^2 - 1$.

Example In the case $\lambda = 0$, $\mu = 1$ a combination of the arguments above yields the following possibilities for (n,k,r,s):

$$(5,2,-\tfrac{1}{2}+\tfrac{1}{2}\sqrt{5},-\tfrac{1}{2}-\tfrac{1}{2}\sqrt{5})\ ,\ (10,3,1,-2)\ ,\ (50,7,2,-3)\ ,\ (3250,56,7,-8)\ .$$

The first three possibilities are uniquely satisfied by the pentagon, the Petersen graph, and the Hoffman-Singleton graph, cf. [11]; in the last case existence is unknown.

5. THE ADJACENCY ALGEBRA

For any two square matrices M and N of size n, the *Schur product* $M \circ N$ is the matrix of size n whose entries are the product of the corresponding entries in M and N. Denoting by $\sum$ the sum of the entries we have

$$\sum M \circ N = \text{trace } MN^t \ .$$

The unit matrix I and the adjacency matrices A and B of any pair of complementary graphs are mutually orthogonal idempotents with respect to Schur multiplication. The vector space $\mathcal{A} = \langle I,A,B\rangle_{\mathbb{R}}$ of dimension 3, consisting of the real linear combinations of I,A,B, is thus closed with respect to Schur multiplication.

From now on, let A and B denote the adjacency matrices of a pair of complementary strongly regular graphs. Then the definition of §3 in terms of matrices implies that $\mathcal{A}$ is closed with respect to ordinary matrix multiplication as well. $\mathcal{A}$ is called the *adjacency algebra* of the strongly regular graphs. We are interested in the *minimal idempotents* E_0,E_1,E_2 with respect to matrix multiplication in $\mathcal{A}$. The following tables list them as linear combinations of the basis $\{I,A,B\}$, and show that they provide another basis for $\mathcal{A}$:

	I	A	B
nE_0	1	1	1
nE_1	f	$f\frac{r}{k}$	$-f\frac{r+1}{\ell}$
nE_2	g	$g\frac{s}{k}$	$-g\frac{s+1}{\ell}$

	E_0	E_1	E_2
I	1	1	1
A	k	r	s
B	ℓ	-r-1	-s-1

Indeed, straightforward calculation by use of the relations between the parameters of §4 shows that the product of the size 3 matrices equals nI_3, and that $E_iE_j = \delta_{ij}E_i$. The bases $\{I =: A_0, A =: A_1, B =: A_2\}$ and $\{E_0, E_1, E_2\}$ of the adjacency algebra A have the following properties. Since A is closed with respect to matrix multiplication and Schur multiplication, we may write, for $i,j \in \{0,1,2\}$:

$$A_iA_j = \sum_{k=0}^{2} p_{ij}^k A_k, \quad E_i \circ E_j = \sum_{k=0}^{2} q_{ij}^k E_k .$$

The *Bose parameters* p_{ij}^k and the *Krein parameters* q_{ij}^k are subject to the following restrictions:

$$p_{ij}^k \in \mathbb{N}, \quad 0 \leq q_{ij}^k .$$

Indeed, p_{ij}^k is the number of the vertices in the relations i and j to any two vertices which are in the relation k. Furthermore, the numbers q_{ij}^k, $k = 0,1,2$, are the eigenvalues of the matrix $E_i \circ E_j$ which, as a principal submatrix of the Kronecker product $E_i \otimes E_j$, is positive semi-definite. From the defining equations for the Krein parameters it follows that

$$nq_{ij}^0 = \sum E_i \circ E_j = \delta_{ij} \operatorname{rank} E_i ,$$

$$nq_{ij}^h q_{hh}^0 = \sum E_h \circ E_i \circ E_j .$$

For $i = 0,1,2$, the idempotent E_i is diagonalized by use of an orthogonal matrix as follows:

$$E_i = [H_i \quad K_i] \begin{bmatrix} I & 0 \\ 0 & 0 \end{bmatrix} \begin{bmatrix} H_i^t \\ K_i^t \end{bmatrix} = H_i H_i^t \quad , \qquad H_i^t H_i = I \quad ,$$

with $H_0 = j/\sqrt{n}$ of size $n \times 1$, H_1 of size $n \times f$, H_2 of size $n \times g$. The n rows of H_i are the coordinates, with respect to an orthonormal basis, of n vectors in the i-th common eigenspace of the matrices in A , and E_i is the Gram matrix of the inner products of these vectors. From the first table above we read that, for $i = 1,2$, the cosines of the angles between the vectors take two values α_i and β_i , namely

$$\alpha_1 = \frac{r}{k} , \quad \beta_1 = -\frac{r+1}{\ell} \quad ; \quad \alpha_2 = \frac{s}{k} , \quad \beta_2 = -\frac{s+1}{\ell} \quad .$$

Hence the n vectors constitute a spherical 2-distance set, X_i say, in the i-th eigenspace.

Dropping the indices, we have for instance in the second eigenspace of dimension g a spherical 2-distance set X consisting of n vectors x whose coordinates are the n rows $(x_1, x_2, \ldots, x_g)$ of the matrix H . From

$$q_{22}^0 = \frac{1}{n} \sum E_2 \circ E_2 = \frac{g}{n} \quad , \qquad q_{22}^2 = \frac{1}{g} \sum E_2 \circ E_2 \circ E_2$$

we find various expressions for q_{22}^2 :

$$g q_{22}^2 = \sum_{x \in X} \sum_{y \in X} (x,y)^3 =$$

$$= \sum_{x \in X} \sum_{y \in X} \sum_{a=1}^{g} x_a y_a \sum_{b=1}^{g} x_b y_b \sum_{c=1}^{g} x_c y_c =$$

$$= \sum_{a=1}^{g} \sum_{b=1}^{g} \sum_{c=1}^{g} \Big(\sum_{x \in X} x_a x_b x_c \Big)^2 \quad .$$

Expressing E_2 in I, A, B we have

$$q_{22}^2 = \frac{g^2}{n^2}\left(1+\frac{s^3}{k^2}-\frac{(s+1)^3}{\ell^2}\right) ,$$

whence, by use of the formula for ℓ of §4:

$$\frac{n^2k^2(r+1)^2}{g^2}\, q_{22}^2 = k^2(r+1)^2\left(1+\frac{s^3}{k^2}-\frac{(s+1)(k+rs)^2}{k^2(r+1)^2}\right) =$$

$$= k^2((r+1)^2-(s+1)) - 2krs(s+1) + s^3(r+1)^2 - r^2s^2(s+1) =$$

$$= (k-s)(k(r^2+2r-s) + s(r^2-2rs-s)) .$$

Therefore, $q_{22}^2 = 0$ is equivalent to any of the following conditions:

$$\sum_{x\in X}\sum_{y\in X}(x,y)^3 = 0 ,$$

$$\sum_{x\in X} x_a x_b x_c = 0 \quad \text{for all} \quad a,b,c \in \{1,2,\dots,g\} ,$$

$$1+\frac{s^3}{k^2}-\frac{(s+1)^3}{\ell^2} = 0 ,$$

$$k(r^2+2r-s) = -s(r^2-2rs-s) .$$

For the significance of the second condition in terms of spherical designs we refer to [7].

6. THE ABSOLUTE BOUND

Let X be a set of n vectors $x,y,\dots$ on the unit sphere Ω_d in $\mathbb{R}^d$. Suppose that the inner products of the vectors of X take two values $\neq 1$, that is,

$$|\{(x,y)\,|\,x \neq y \in X\}| = 2 .$$

Then the cardinality $n = |X|$ is bounded in terms of the

dimension d by the *absolute bound*

$$n \leq \tfrac{1}{2}d(d+3)$$

This is proved as follows, cf. [13]. Let α and β be the admissible inner products of X. For each $y \in X$ we define

$$F_y(\xi) := \frac{((y,\xi)-\alpha)((y,\xi)-\beta)}{(1-\alpha)(1-\beta)} \quad , \qquad \xi \in \Omega_d \quad .$$

These are n polynomials of degree ≤ 2 in the coordinates $\xi_1,\ldots,\xi_d$ of ξ. These polynomials are independent since for all $x,y \in X$

$$F_y(x) = \delta_{y,x} \quad .$$

The linear space of all polynomials of degree ≤ 2 has a basis consisting of the $\tfrac{1}{2}d(d+1) + d$ polynomials

$$\xi_1^2, \xi_1\xi_2, \xi_1\xi_3, \ldots, \xi_d^2, \xi_1, \xi_2, \ldots, \xi_d \quad .$$

Notice that the constant polynomial is dependent on these since

$$1 = \xi_1^2 + \ldots + \xi_d^2 \quad .$$

This implies the absolute bound. If only supplementary angles are admitted then $\beta = -\alpha$ in the above, and the polynomials $F_y(\xi)$ are homogeneous. Therefore, in this case we arrive at the sharper bound

$$n \leq \tfrac{1}{2}d(d+1) \quad .$$

For a spherical 2-distance set the absolute bound $\tfrac{1}{2}d(d+3)$ is attained if and only if the n independent polynomials

F_y, $y \in X$, provide a basis for the linear space of all polynomials of degree ≤ 2. Then in particular, for any $x \in X$, we have

$$(x,\xi)^2 = \sum_{y \in X} (x,y)^2 F_y(\xi) \quad , \qquad \xi \in \Omega_d \quad .$$

The terms of degree one yield

$$0 = (\alpha+\beta) \sum_{y \in X} (x,y)^2 (y,\xi) \quad , \qquad \xi \in \Omega_d \quad .$$

Since $\alpha + \beta = 0$ is impossible we arrive at

$$\sum_{y \in X} (x,y)^3 = 0 \quad , \qquad \text{for all} \quad x \in X \quad .$$

Now let A be the adjacency matrix of a strongly regular graph. In §5 we observed that, for $i = 1,2$, the i-th eigenspace of A contains a set X_i of n vectors with Gram matrix E_i, constituting a spherical 2-distance set. As a consequence of the absolute bound, the parameters of a strongly regular graph satisfy

$$n \leq \tfrac{1}{2}f(f+3) \quad , \qquad n \leq \tfrac{1}{2}g(g+3) \quad .$$

Moreover, as a consequence of the above and §5 we have:

$$n = \tfrac{1}{2}f(f+3) \quad \text{implies} \quad q_{11}^1 = 0 \quad , \qquad \text{and}$$

$$n = \tfrac{1}{2}g(g+3) \quad \text{implies} \quad q_{22}^2 = 0 \quad .$$

We conclude this section with a few remarks on old and new results which follow from the inequalities met so far. A. Neumaier [14], by combining inequalities of several types, proves that

$$\mu \le s^3(2s+3) \ , \quad \text{with equality iff} \quad n = \tfrac{1}{2}f(f+3) \ .$$

Thus he makes precise the result by Hoffman that μ is bounded in terms of the smallest eigenvalue s, which was used by Sims in his classification of strongly regular graphs, cf. [16] theorems 13 and 14. In fact,

$$f = \frac{-(s+1)(\mu-rs)(\mu-r(s+1))}{\mu(r-s)} \ ,$$

implies that $r-s$ divides $-(s+1)(\mu-s^2)(\mu-s(s+1))$, hence also r is bounded in terms of s, provided $\mu \ne s^2$ and $\mu \ne s(s+1)$. This leads to the important theorem that a strongly regular graph with given s is of Steiner or Latin square type or is one of finitely many exceptions (Sims, Ray-Chaudhuri [15], Neumaier [14]).

Neumaier [14] also finds a new inequality (the claw bound) which, due to a remark by A.E. Brouwer, takes the form:

$$2(r+1) \le s(s+1)(\mu+1) \ , \quad \text{if} \quad \mu \ne s^2 \ \text{and} \ \mu \ne s(s+1) \ .$$

7. EXTREMAL STRONGLY REGULAR GRAPHS

The pentagon is strongly regular with

$$n = \tfrac{1}{2}f(f+3) = \tfrac{1}{2}g(g+3) \ , \qquad q_{11}^1 = q_{22}^2 = 0 \ .$$

We claim that the pentagon is the only such graph. For other strongly regular graphs we claim the following:

(i) $q_{22}^2 = 0$ holds if and only if

$$n = \frac{-2(r-s)^2(r^2+2rs+2r+s)}{(r-s)^2 - r^2(r+1)^2} \ , \quad k = \frac{-s(r^2-2rs-s)}{r^2+2r-s} \ ;$$

$$r,s \in \mathbb{Z} \ , \quad r(r+3) \le r-s \le r(r+1)(2r+1) \ ,$$

and some divisibility conditions.

(ii) $n = \frac{1}{2}g(g+3)$ holds if and only if

$$n = (2r+1)^2(2r^2+2r-1),\ k = 2r^3(2r+3),\ s = -r^2(2r+3),$$
$$r \in \mathbb{Z}^+ .$$

(iii) $q_{22}^2 = 0 = p_{11}^1$ holds if and only if

$$n = r^2(r+3)^2,\ k = r^3 + 3r^2 + r,\ s = -r^2 - 2r,\ r \in \mathbb{Z}^+ .$$

To prove this, we recall from §5 that $q_{22}^2 = 0$ is equivalent to

$$k(r^2+2r-s) = -s(r^2-2rs-s) \quad ,$$

whence the formulae for k and n. In the case $q_{22}^2 = 0$ the conditions $0 \le p_{11}^1$ and $n \le \frac{1}{2}g(g+3)$ read

$$r(r+3) \le r - s \le r(r+1)(2r+1) \quad .$$

Since $r + s = -1$ only yields the pentagon, the results of §4 imply $r, s \in \mathbb{Z}$. Since $n = \frac{1}{2}g(g+3)$ implies $q_{22}^2 = 0$ by §6, we have $n = \frac{1}{2}g(g+3)$ iff $r - s = r(r+1)(2r+1)$, and $q_{22}^2 = p_{11}^1 = 0$ iff $r - s = r(r+3)$. Now the values for the parameters follow and our claims are proved.

For $r = 1$ and $r = 2$ there remain the following possibilities for $q_{22}^2 = 0$, each of which is satisfied by a unique well-known graph:

$$(n,k,r,s) = (27,10,1,-5),(275,112,2,-28) \text{ for } n = \tfrac{1}{2}g(g+3) ,$$

$$(n,k,r,s) = (16,5,1,-3),(100,22,2,-8) \text{ for } p_{11}^1 = 0 \quad ,$$

$$(n,k,r,s) = (112,30,2,-10),(162,56,2,-16)$$

These graphs are the complement of the Schläfli graph (cf. §9) for $n = 27$, the Clebsch graph for $n = 16$, the Higman-Sims graph for $n = 100$, and the McLaughlin graph for $n = 275$ whose subconstituents with respect to any vertex are the 112-graph and the 162-graph, cf. [7], [9].

The present extremal strongly regular graphs admit a graph-theoretical interpretation in terms of their subconstituents with respect to any vertex. In terms of the notation at the end of §5, selection of any vertex in a strongly regular graph amounts to selection of any $x \in X$ in the second eigenspace, say. If the orthonormal basis of the eigenspace is chosen to contain x, then the adjacency matrices A and B and the coordinate matrix H of the vectors from X take the following form:

$$A = \begin{bmatrix} 0 & j^t & 0^t \\ j & A' & M \\ 0 & M^t & A'' \end{bmatrix}, \quad B = \begin{bmatrix} 0 & 0^t & j^t \\ 0 & B' & N \\ j & N^t & B'' \end{bmatrix}, \quad \sqrt{\frac{n}{g}}H = \begin{bmatrix} 1 & 0^t \\ \alpha j & K \\ \beta j & L \end{bmatrix}.$$

From the definition of H it follows that

$$I + \alpha A' + \beta B' = \alpha^2 J + KK^t \;, \quad I + A' + B' = J \;,$$

$$I + \alpha A'' + \beta B'' = \beta^2 J + LL^t \;, \quad I + A'' + B'' = J \;,$$

$$JK = 0 \;, \quad JL = 0 \;, \quad K^t K + L^t L = \frac{n}{g} I \;.$$

If $q_{22}^2 = 0$ then the subconstituents, with the adjacency matrices A', A'', B', B'', are strongly regular, possibly with $\mu = 0$. Indeed, $q_{22}^2 = 0$ implies by §5 that

$$\sum_{y \in X} y_1 y_\mu y_\nu = 0 \quad ,$$

for all $\mu,\nu \in \{2,\dots,g\}$, hence

$$\alpha K^t K + \beta L^t L = 0 \quad .$$

It follows that K^tK and L^tL are multiples of I , hence A',A'',B',B'' have the right number of eigenvalues. It turns out, cf. [7], that if $q_{22}^2 = 0$ and $g < f$ then the orders and the eigenvalues of the subconstituents are as follows:

$$A' : n' = \frac{-s(r^2-2rs-s)}{r^2+2r-s}, \quad k' = \frac{r(s+1)(r^2+2r+s)}{r^2+2r-s},$$

$$r' = r, \quad s' = \tfrac{1}{2}(r^2+2r+s),$$

$$A'' : n'' = \frac{(s+1)(r^2-2rs-s)}{r^2+s}, \quad k'' = -rs, \; r'' = r, \; s'' = \tfrac{1}{2}(s-r^2)$$

If $n = \frac{1}{2}g(g+3)$ then $-s = r^2(2r+3)$, hence

$$A' : n' = 2r^3(2r+3), \; k' = r(2r-1)(r^2+r-1), \; r' = r,$$

$$s' = -r(r^2+r-1),$$

$$A'' : n'' = 2(r+1)^3(2r-1), \; k'' = r^3(2r+3), \; r'' = r,$$

$$s'' = -r^2(r+2) .$$

If $q_{22}^2 = 0 = p_{11}^1$ and $g < f$ then $-s = 2r + r^2$, hence

$$A' : n' = r^3 + 3r^2 + r, \quad k' = 0, \quad r' = 0, \quad s' = 0,$$

$$A'' : n'' = (r^2+2r-1)(r^2+3r+1), \quad k'' = r^3 + 3r^2, \quad r'' = r,$$

$$s'' = -r^2 - r .$$

Conversely, what are the consequences for the whole graph of conditions imposed on the subconstituents? In the case that A is strongly regular with $q_{22}^{2} = 0$ and $g < f$, straightforward calculation by use of the eigenvalues for A' and A'' given above leads to the following:

A'' satisfies $q_{22}^{2} = 0$ iff A satisfies $n = \frac{1}{2}g(g+3)$;

A' satisfies $q_{22}^{2} = 0$ iff A satisfies $n = \frac{1}{2}g(g+3)$ or $p_{11}^{1} = 0$.

We refer to [7] for the proof of the following result. A strongly regular graph, not of pseudo or negative Latin square type, satisfies $q_{11}^{1} = 0$ or $q_{22}^{2} = 0$ if both subconstituents with respect to some vertex are strongly regular.

8. SWITCHING

Let A and B be the adjacency matrices of a pair of complementary strongly regular graphs, and let their eigenvalues be as in the table of §5. Then it follows from the relations between the parameters of §4 that $B - A$ satisfies

$$(B - A + (2r+1)I)(B - A + (2s+1)I) = (n + 4rs + 2r + 2s)J ,$$

$$(B - A)J = (\ell - k)J .$$

We consider the special case when

$$n + 4rs + 2r + 2s = 0 .$$

Then the matrix $B - A$ satisfies

$$(B - A + (2r+1)I)(B - A + (2s+1)I) = 0 ,$$

and has exactly two eigenvalues $-2s-1$ and $-2r-1$, of multiplicities g and $f+1$ or $g+1$ and f, depending on whether the eigenvalue $\ell - k$ belonging to the all-one eigenvector equals $-2r-1$ or $-2s-1$. We now multiply certain rows and the corresponding columns of $B - A$ by -1. We write the resulting matrix as $B' - A'$, where B' and $-A'$ are the matrices of the ones and of the minus ones, respectively. The (1,0) matrices A' and B' are interpreted as the adjacency matrices of a pair of complementary graphs. Notice that the graphs indicated by A and by A' may be completely different; however, the matrices $B - A$ and $B' - A'$ have the same eigenvalues. We say that the graph A is *switched* into the graph A'. If switching is performed in such a way that the resulting matrix $B' - A'$ has constant row sums, then the new graph A' again is strongly regular, either with different parameters, or with the same parameters but nonisomorphic to the old graph A, or isomorphic to the old graph A. We can also switch so as to make any one vertex into an isolated vertex; then the new graph (after removal of the isolated vertex) is strongly regular with $k = 2\mu$. Conversely, any such graph when extended by an isolated vertex yields a graph whose $B - A$ has only two eigenvalues.

Thus, given a strongly regular graph with $n + 4rs + 2r + 2s = 0$, we have a possibility to construct new strongly regular graphs on n and on $n - 1$ vertices. In addition, given a strongly regular graph with $k = 2\mu$, we have a possibility to construct new strongly regular graphs on $n + 1$ and on n vertices. An example of this last phenomenon is provided by the strongly regular graphs on $\frac{1}{2}g(g+3)$ vertices of §7.

The next section illustrates the above in two explicit examples. For further such examples the reader is referred to [19].

9. TWO EXAMPLES

The triangular graph $T(8)$ satisfies

$$(B-A-3I)(B-A+9I)=0\ ,\quad (B-A)J=3J\ .$$

We describe its vertices by the 28 pairs from the set $\{a,b,c,d,e,f,g,h\}$. Switching with respect to the 12 vertices

$$\{a,i\},\{b,i\}\ ,\quad \text{where}\quad i\in\{c,d,e,f,g,h\}$$

isolates the vertex $\{a,b\}$. The remaining graph is the well-known Schläfli graph on 27 vertices with the eigenvalues and multiplicities

$$k=16\ ,\quad r=4\ ,\quad s=-2\ ,\quad f=6\ ,\quad g=20\ .$$

The triangular graph $T(8)$ itself has the parameters

$$n=28\ ,\ k=12\ ,\ r=4\ ,\ s=-2\ ,\ f=7\ ,\ g=20\ .$$

Three nonisomorphic graphs with the same parameters are obtained from $T(8)$ by switching

with respect to $\{a,b\},\{c,d\},\{e,f\},\{g,h\}$;

with respect to $\{a,b\},\{b,c\},\{c,d\},\{d,e\},\{e,f\},\{f,g\},$
$\{g,h\},\{h,a\}$;

with respect to $\{a,b\},\{b,c\},\{c,a\},\{d,e\},\{e,f\},\{f,g\},$
$\{g,h\},\{h,d\}$.

These are the well-known Chang graphs. Together with $T(8)$ they provide the only graphs with these parameters, cf. [18]. There do not exist strongly regular graphs with

$$n = 28\,,\ k = 18\,,\ r = 4\,,\ s = -2\,,\ f = 6\,,\ g = 21\,,$$

$$\ell - k = -(2r+1) = -9\ .$$

Indeed, these parameters violate both the condition $q_{11}^{1} \geq 0$ of §5 and the absolute bound $n \leq \frac{1}{2}f(f+3)$ of §6.

Our second example concerns the strongly regular graph with $n = 35$, $k = 18$, $r = 3$, $s = -3$, which is constructed from one of the 80 Steiner triple systems on 15 symbols by the method of the example in §3. The vertices are the 35 lines of $PG(3,2)$, two vertices being adjacent whenever the corresponding lines intersect. The adjacency matrices A and B of this graph and its complement satisfy

$$(B - A - 5I)(B - A + 7I) = -2J\ .$$

Extending the graph by an isolated vertex we obtain a non-regular graph. We claim that this graph can be switched into a regular graph with $\ell - k = -7$, and into a regular graph with $\ell - k = 5$, so as to obtain strongly regular graphs with

$$n = 36\,,\ k = 21\,,\ r = 3\,,\ s = -3 \quad \text{and}$$

$$n = 36\,,\ k = 15\,,\ r = 3\,,\ s = -3\ .$$

Indeed, in the first case the switching set consists of all lines of $PG(3,2)$ except for the 7 lines of any plane and the 7 lines through any point not in the plane. This unique possibility yields a rank 3 graph which is a subconstituent of the rank 3 graph on 100 vertices having as its automorphism group the Hall-Janko group. In the second case the switching set consists of 15 lines of PG(3,2) such that each of the 15 points is on 3 lines. One such set is obtained by taking 6 spreads (of 5 skew lines), each pair of spreads having one line in common. The graph thus obtained is a rank 3 graph having as its automorphism group the orthogonal group $O^{-}(6,2)$. There are two further nonisomorphic strongly regular graphs with $k = 15$ obtained in this way from the lines of PG(3,2).

By starting from one of the 80 Steiner triple systems on 15 symbols, or from one of the 12 Latin squares on 6 symbols, the following numbers of nonisomorphic strongly regular graphs with

$$(n,k,r,s) = (36,21,3,-3),(36,15,3,-3),(35,18,3,-3)$$

are obtained:

	from STS	from LS	total
(36,21,3,-3)	57	48	105
(36,15,3,-3)	16111	337	16448
(35,18,3,-3)	1815	38	1853

For details we refer to [3], [4].

Techn. Univ. Eindhoven,
The Netherlands.

REFERENCES

[1] Bose, R.C. (1963). Strongly regular graphs, partial geometries, and partially balanced designs. *Pacific J.*

Math. 13, 389-419.

[2] Bose, R.C. and Mesner, D.M. (1959). On linear associative algebras corresponding to association schemes of partially balanced designs. *Ann. Math. Statist.* 30, 21-38.

[3] Bussemaker, F.C. and Seidel, J.J. (1970). Symmetric Hadamard matrices of order 36. *Ann. N.Y. Acad. Sci.* 175, 66-79; Report Techn. Univ. Eindhoven 70-WSK-02.

[4] Bussemaker, F.C., Mathon, R. and Seidel, J.J. Tables of two-graphs. Report Techn. Univ. Eindhoven (in preparation).

[5] Cameron, P.J. and van Lint, J.H. (1975). Graph theory, coding theory and block designs. *London Math. Soc.* Lecture Notes 19.

[6] Cameron, P.J. (1978). Strongly regular graphs. In *Selected Topics in Graph Theory*, eds. L.W. Beineke and R.J. Wilson. Academic Press.

[7] Cameron, P.J., Goethals, J.M. and Seidel, J.J. (1978) Strongly regular graphs having strongly regular subconstituents. *J. Algebra* 55, 257-280.

[8] Goethals, J.M. and Seidel, J.J. (1970). Strongly regular graphs derived from combinatorial designs. *Canad. J. Math.* 22, 597-614.

[9] Goethals, J.M. and Seidel, J.J. (1975). The regular two-graph on 276 vertices. *Discrete Math.* 12, 143-158.

[10] Higman, D.G. (1971). A survey of some questions and results about rank 3 permutation groups. *Actes Congrès Intern. Math. Nice 1970*, I, 361-365. Gauthier-Villars.

[11] Hoffman, A.J. (1975). Eigenvalues of graphs, pp.225-245. In *Studies in Graph Theory, Part II*, ed. D.R. Fulkerson, Math. Assoc. Amer.

[12] Hubaut, X. (1975). Strongly regular graphs. *Discrete Math.* 13, 357-381.

[13] Koornwinder, T.H. (1976). A note on the absolute bound for systems of lines. *Proc. Kon. Nederl. Akad. Wet. A* 79 (= *Indag. Math.* 38), 152-153.

[14] Neumaier, A. (1978). Lecture Techn. Univ. Eindhoven. September 1978.

[15] Ray-Chaudhuri, D.K. (1976). Uniqueness of association schemes. *Proc. Intern. Colloqu. Teorie Combinatorie* (Roma 1973), Tomo II, 465-479, *Accad. Naz. Lincei.*

[16] Ray-Chaudhuri, D.K. (1977). Geometric incidence structures, pp.87-116. In *Proc. 6th British Combinatorial Conference*, ed. P.J. Cameron, Academic Press.

[17] Scott, L.L. (1973). A condition on Higman's parameters. *Amer. Math. Soc. Notices* 701-02-45.

[18] Seidel, J.J. (1967). Strongly regular graphs of L_2-type and of triangular type. *Proc. Kon. Nederl. Akad. Wet. A* 70 (= *Indag. Math.* 29), 188-196.

[19] Seidel, J.J. (1969). Strongly regular graphs, pp.185-197. In *Progress in Combinatorics*, ed. W. Tutte, Academic Press.

[20] Smith, M.C. (1975). On rank 3 permutation groups. *J. Algebra* 33, 22-42.

7 · Geometries in finite projective and affine spaces

J.A. THAS

1. SOME INTERESTING FINITE GEOMETRIES

1.1. Generalized quadrangles

A (finite) generalized quadrangle is an incidence structure $S=(P,B,I)$, with a symmetric incidence relation satisfying the following axioms

(i) each point is incident with $1+t$ ($t\geq 1$) lines and two distinct points are incident with at most one line

(ii) each line is incident with $1+s$ ($s\geq 1$) points and two distinct lines are incident with at most one point;

(iii) if x is a point and L is a line not incident with x, then there is a unique point x' and a unique line L' for which $xIL'Ix'IL$.

We have $|P|=v=(1+s)(1+st)$ and $|B|=b=(1+t)(1+st)$. D.G. Higman proved that for $s>1$ and $t>1$, there holds $t\leq s^2$ and dually $s\leq t^2$ [27]. Moreover $s+t \mid st(1+s)(1+t)$ [27], [8].

Examples

(1) The classical generalized quadrangles. (a) We consider a non-singular hyperquadric Q of index 2 of the projective space $PG(d,q)$, with $d=3,4$ or 5. Then the points of Q together with the lines of Q (which are the subspaces of maximal dimension on Q) form a generalized quadrangle $Q(d,q)$ with parameters [22]

$s=q$, $t=1$, $v=(q+1)^2$, $b=2(q+1)$, when $d=3$;

$s=t=q$, $v=b=(q+1)(q^2+1)$, when $d=4$;

$s=q$, $t=q^2$, $v=(q+1)(q^3+1)$, $b=(q^2+1)(q^3+1)$, when $d=5$.

(b) Let H be a non-singular hermitian variety of the projective space $PG(d,q^2)$, $d=3$ or 4. Then the points of H together with the lines on H form a generalized quadrangle $H(d,q^2)$ with parameters [22]

$s=q^2$, $t=q$, $v=(q^2+1)(q^3+1)$, $b=(q+1)(q^3+1)$, when $d=3$;

$s=q^2$, $t=q^3$, $v=(q^2+1)(q^5+1)$, $b=(q^3+1)(q^5+1)$, when $d=4$.

(c) The points of $PG(3,q)$, together with the totally isotropic lines with respect to a symplectic polarity, form a generalized quadrangle $W(3,q)$ with parameters $s=t=q$, $v=b=(q+1)(q^2+1)$ [22].

All these generalized quadrangles (all of which are associated with classical simple groups) are due to J. Tits.

(2) Two non-classical generalized quadrangles. (a) Let O be a complete oval of the projective plane $PG(2,q)$, q even, and let $PG(2,q)$ be embedded as a plane H in $PG(3,q)=V$. Define points of the generalized quadrangle as the points of V-H. Lines of the quadrangle are the lines of V which are not contained in H and meet O (necessarily in a unique point). The incidence is that of V. The parameters are $s=q-1$, $t=q+1$, $v=q^3$, $b=q^2(q+2)$. This quadrangle was first discovered by R. Ahrens and G. Szekeres [1] and independently by M. Hall, Jr. [26].
It will be denoted by $T_2^*(O)$.

(b) Consider the quadrangle $W(3,q)$, and let x be one of its points. Call $PG(2,q)$ the polar plane of x with respect to the symplectic polarity defining $W(3,q)$. Then a new quadrangle $W^*(3,q)$ is constructed as follows :
points of $W^*(3,q)$ are the points of $W(3,q)$ which are not

in PG(2,q); lines of $W^*(3,q)$ are the lines of W(3,q) which do not contain x, and also the lines of PG(3,q) which contain x and are not contained in PG(2,q); the incidence is the natural one. This quadrangle with parameters $s=q-1$, $t=q+1$, $v=q^3$, $b=q^2(q+2)$ was first discovered by R. Ahrens and G. Szekeres [1].

Remark. The classical generalized quadrangles are embedded in a projective space, and the two given non-classical examples (there are other examples) are embedded in the affine space AG(3,q).

Literature. Generalized quadrangles were introduced in 1959 by J. Tits [53]. For a survey on the subject we refer to S.E. Payne [39], J.A. Thas and S.E. Payne [52], and J.A. Thas [49].

1.2. Partial geometries

A (finite) partial geometry is an incidence structure $S=(P,B,I)$, with a symmetric incidence relation satisfying the following axioms

(i) each point is incident with $1+t$ ($t \geq 1$) lines and two distinct points are incident with at most one line;

(ii) each line is incident with $1+s$ ($s \geq 1$) points and two distinct lines are incident with at most one point;

(iii) if x is a point and L is a line not incident with x, then there are exactly α ($\alpha \geq 1$) points $x_1,\dots,x_\alpha$ and α lines $L_1,\dots,L_\alpha$ such that $x I L_i I x_i I L$, $i=1,2,\dots,\alpha$.

We have $|P|=v=(1+s)(st+\alpha)/\alpha$ and $|B|=b=(1+t)(st+\alpha)/\alpha$. There holds $\alpha(s+t+1-\alpha) \mid st(s+1)(t+1)$ [27], [8], $(t+1-2\alpha)s \leq (t+1-\alpha)^2(t-1)$ [12], and dually $(s+1-2\alpha)t \leq$

$\leq (s+1-\alpha)^2 (s-1)$. The partial geometries with $\alpha=1$ are of course the generalized quadrangles.

The four classes of partial geometries

(a) The partial geometries with $\alpha=s+1$ or, dually, $\alpha=t+1$. A partial geometry with $\alpha=s+1$ is the same as a 2-$(v,s+1,1)$ design.

(b) The partial geometries with $\alpha=s$, or, dually, $\alpha=t$. A partial geometry with $\alpha=t$ is the same as a net of order $s+1$ and degree $t+1$ (or deficiency $t-s+1$).

(c) The generalized quadrangles.

(d) The partial geometries with $1<\alpha<\min(s,t)$.

Examples

(1) Partial geometries arising from maximal arcs. In a finite projective plane of order q, any non-void set of l points may be described as a $\{l;n\}$-arc, where $n (n\neq 0)$ is the greatest number of collinear points in the set. For given q and n $(n\neq 0)$, l can never exceed $nq-q+n$, and an arc with that number of points will be called a maximal arc [3]. Equivalently, a maximal arc may be defined as a non-void set of points meeting every line in just n points or in none at all.

If K is a $\{qn-q+n;n\}$-arc (i.e. a maximal arc) of a projective plane π of order q, where $n\leq q$, then the set $K'=\{\text{lines } L \text{ of } \pi \,\|\, L\cap K=\phi\}$ is a $\{q(q-n+1)/n;\ q/n\}$-arc (i.e. a maximal arc) of the dual projective plane π^* of π. It follows that, if the desarguesian projective plane $PG(2,q)$ contains a $\{qn-q+n;n\}$-arc, $n\leq q$, then it also contains a $\{q(q-n+1)/n;\ q/n\}$-arc.

From the preceding there follows that a necessary condition for the existence of a maximal arc (as a proper subset of a given plane) is that n should be a factor of q. R.H.F. Denniston [23] has proved that the condition does suffice in the case of any desarguesian plane of order 2^h. This is not the case for odd order desarguesian planes. A. Cossu [13] has proved that there is no $\{21;3\}$-arc in PG(2,9). In J.A. Thas [47] appears the following more general result : in PG(2,q), $q=3^h$ and $h>1$, there are no $\{2q+3;3\}$-arcs and no $\{q(q-2)/3;q/3\}$-arcs. We conjecture that in PG(2,q), q odd, there are no $\{qn-q+n;n\}$-arcs with $1<n<q$ (i.e. the only maximal arcs of PG(2,q), q odd, are PG(2,q), AG(2,q) and the points of PG(2,q)).

Let K be a $\{qn-q+n;n\}$-arc, $1<n<q$, of a projective plane π (not necessarily desarguesian) of order q. Define points of the partial geometry S(K) as the points of π which are not contained in K. Lines of S(K) are the lines of π which are incident with n points of K. The incidence is that of π. Then S(K) is a partial geometry with parameters $s=q-n$, $t=q-q/n$, $\alpha=q-q/n-n+1$ [48] , [55] .

In particular, let K be a $\{qn-q+n;n\}$-arc, $1<n<q$, of PG(2,q).Then a second partial geometry $T_2^*(K)$ is defined as follows. Let PG(2,q) be embedded as a plane H in PG(3,q)=V. Define points of the geometry as the points of V-H. Lines of $T_2^*(K)$ are the lines of V which are not contained in H and meet K. The incidence is that of V. Then $T_2^*(K)$ is a partial geometry with parameters $s=q-1$, $t=qn-q+n-1$, $\alpha=n-1$ [48] .

As the existence of the $\{qn-q+n;n\}$-arc K in PG(2,q) implies the existence of a $\{q(q-n+1)/n;q/n\}$-arc K' in PG(2,q), there also exists the partial geometry $T_2^*(K')$ with parameters $s=q-1$, $t=(q^2-qn+q-n)/n$, $\alpha=(q-n)/n$ [48] .

Any known partial geometry with $1<\alpha<\min(s,t)$ is a $S(K)$ or a $T_2^*(K)$.

We remarked that there exist $\{2^{m+h}-2^h+2^m ; 2^m\}$-arcs in $PG(2,2^h)$ $(0\leq m\leq h)$, and hence there exist partial geometries with parameters

(a) $s=2^h-2^m$, $t=2^h-2^{h-m}$, $\alpha=(2^m-1)(2^{h-m}-1)$ $(0<m<h)$, and

(b) $s=2^h-1$, $t=2^{h+m}-2^h+2^m-1$, $\alpha=2^m-1$ $(0<m<h)$.

A partial geometry (a) is a generalized quadrangle iff $h=2$ and $m=1$ (then $s=t=2$, $v=b=15$). A partial geometry (b) is a generalized quadrangle iff K is a complete oval O.

(2) <u>The partial geometry</u> H_q^n. P is the set of all points of $PG(n,q)$ which are not contained in a given subspace $PG(n-2,q)$ $(n\geq 3)$; B is the set of all lines of $PG(n,q)$ which do not have a point in common with $PG(n-2,q)$; I is the natural incidence relation. Then (P,B,I) is a partial geometry with parameters $s=q$, $t=q^{n-1}-1$, $\alpha=q$. This well-known dual net is denoted by H_q^n. It is the only known partial geometry with $\alpha\notin\{1,s+1, t+1\}$ and satisfying the axiom of Pasch [50]. In [50] F. De Clerck and J.A. Thas prove that H_s^n is the only dual net of order $s+1$ (≥ 3) and deficiency $t-s+1$ (>0) which satisfies the axiom of Pasch (independently C.C. Sims proved the same result, but with the restriction $t+1>s^2$).

<u>Remark</u>. The geometries H_q^n are embedded in the projective space $PG(n,q)$, and the geometries $T_2^*(K)$ are embedded in the affine space $AG(3,q)$.

<u>Literature</u>. Partial geometries were introduced by R.C. Bose [7]. For a survey on the subject we refer to J.A. Thas [49] and F. De Clerck [20].

1.3. Semi partial geometries

A (finite) semi partial geometry is an incidence structure $S=(P,B,I)$, with a symmetric incidence relation satisfying the following axioms

(i) each point is incident with $1+t$ $(t\geq 1)$ lines and two distinct points are incident with at most one line;

(ii) each line is incident with $1+s$ $(s\geq 1)$ points and two distinct lines are incident with at most one point;

(iii) if two points are not collinear, then there are μ $(\mu>0)$ points collinear with both;

(iv) if a point x and a line L are not incident, then there are 0 or α $(\alpha\geq 1)$ points which are collinear with x and incident with L (i.e. there are 0 or α points x_i and respectively 0 or α lines L_i such that $xIL_i\,Ix_i\,IL$).

We have $|P|=v=1+(t+1)s(1+t(s-\alpha+1)/\mu)$, and $v(t+1)=b(s+1)$ with $|B|=b$. There holds $\alpha^2\leq\mu\leq(t+1)\alpha$, $\mu\mid(t+1)st(s+1-\alpha)$, $\alpha\mid st(t+1)$, $\alpha\mid st(s+1)$, $\alpha\mid\mu$, $\alpha^2\mid\mu st$, $\alpha^2\mid t((t+1)\alpha-\mu)$, and $b\geq v$ if $\mu\neq(t+1)\alpha$ [18]. Moreover $D=(t(\alpha-1)+s-1-\mu)^2+4((t+1)s-\mu)$ is a square, except in the case $\mu=s=t=\alpha=1$ where $D=5$ (and then S is a pentagon), and $((t+1)s+(v-1)\times(t(\alpha-1)+s-1-\mu+\sqrt{D})/2)/\sqrt{D}$ is an integer [18].

A semi partial geometry with $\alpha=1$ is called a partial quadrangle. Partial quadrangles were introduced and studied by P.J. Cameron [11]. A semi partial geometry is a partial geometry iff $\mu=(t+1)\alpha$. The dual of a semi partial geometry S is again a semi partial geometry iff $s=t$ or S is partial geometry [16].

If we write "$\longrightarrow$" for "generalizes to", then we have the scheme

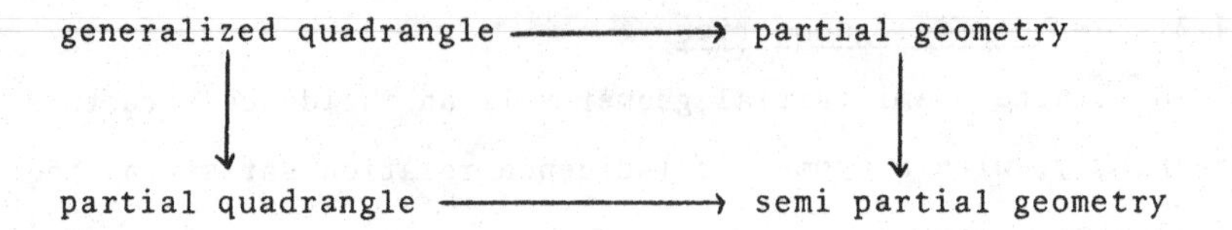

Examples

(1) Let B' (resp. U) be a Baer subplane (resp. unital) of the projective plane PG(2,q), with q a square. Let PG(2,q) be embedded in the projective space PG(3,q). Call P the set of the points of PG(3,q) which do not belong to PG(2,q), B the set of lines of PG(3,q) which do not belong to PG(2,q) and contain exactly one point of B' (resp. U), and I the natural incidence. Then (P,B,I) is a semi partial geometry with parameters $t=q+\sqrt{q}$, $s=q-1$, $\alpha=\sqrt{q}$ and $\mu=q+\sqrt{q}$ (resp. $t=q\sqrt{q}$, $s=q-1$, $\alpha=\sqrt{q}$ and $\mu=q(q-1)$) [18]. This geometry will be denoted by $T_2^*(B')$ (resp. $T_2^*(U)$).

(b) Consider an ovoid O of PG(3,q). Let PG(3,q) be embedded in the projective space PG(4,q). If we define P to be the set of points of PG(4,q) which do not belong to PG(3,q), B to be the set of lines L of PG(4,q) which do not belong to PG(3,q) and contain exactly one point of O, and I to be the incidence relation of PG(4,q), then (P,B,I) is a semi partial geometry with parameters $t=q^2$, $s=q-1$, $\alpha=1$ and $\mu=q(q-1)$. This partial quadrangle, denoted by $T_3^*(O)$, is due to P.J. Cameron [11].

(3) Consider a (d-2)-dimensional subspace PG(d-2,q) of PG(d,q) ($d\geq 3$). Call P the set of all lines of PG(d,q) which have no point in common with PG(d-2,q). Call B the set of all planes of PG(d,q) which have exactly one point in common with PG(d-2,q). An element x of P is incident with an element L of B iff x is contained in

L. Then (P,B,I) is a semi partial geometry with parameters $t=\frac{q^{d-1}-1}{q-1}-1$, $s=q^2-1$, $\alpha=q$ and $\mu=q(q+1)$ [18].

For more information about this geometry we refer to I. Debroey [15], [16].

(4) Let P be the point set of $PG(n,q)$, n odd and $n\geq 3$, let B be the set of all lines which are not totally isotropic for a given symplectic polarity of $PG(n,q)$, and let I be the natural incidence. Then (P,B,I) is a semi partial geometry with parameters $t=q^{n-1}-1$, $s=q$, $\alpha=q$, $\mu=q^{n-1}(q-1)$. It will be denoted by $\bar{W}(n,q)$ [18].

Remarks. 1. Several other examples of semi partial geometries, which are not partial geometries, are known [16].

2. The geometries $T_2^*(B')$ and $T_2^*(U)$ are embedded in $AG(3,q)$; the geometry $T_3^*(O)$ is embedded in $AG(4,q)$. The semi partial geometry $\bar{W}(n,q)$ is embedded in $PG(n,q)$.

Literature. Semi partial geometries were introduced by I. Debroey and J.A. Thas [18]. Further we refer to I. Debroey [16], which is a survey on the subject.

1.4. Partial three-spaces

The concept of partial three-space is due to R. Laskar and J. Dunbar [33], and is defined as follows.

A partial three-space S is a system of points, lines and planes, together with an incidence relation for which the following conditions are satisfied :

(i) If a point p is incident with a line L, and L is incident with a plane π, then p is incident with π.

(ii) (a) A pair of distinct planes is incident with at most one line.

(b) A pair of planes if not incident with a line is incident with at most one point.

(iii) The set of points and lines incident with a plane forms a partial geometry with parameters s,t and α .

(iv) The set of lines and planes incident with a point p forms a partial geometry with parameters s^* , t and α^*, where the points and lines of the geometry are the planes and lines through p respectively and incidence is that of S.

(v) Given a plane π, and a line L not incident with π, π and L not intersecting in a point, there exist exactly u planes through L intersecting π in a line and exactly w-u planes through L intersecting π in a point but not in a line.

(vi) Given a point p and a line L, p and L not coplanar, there exist exactly u^* points on L which are collinear with p, and w^*-u^* points on L coplanar but not collinear with p.

(vii) Given a point p and a plane π not containing p, then there exist exactly x planes through p intersecting π in a line.

Given a partial three-space S, a dual partial three-space is obtained by interchanging points and planes. Also, if two points of S are called first associates if they are collinear, second associates if they are coplanar but not collinear and third associates otherwise, then it is shown in [33] that the points of S form a 3-class association scheme.

Examples

(1) 3-nets. A 3-net, due to R. Laskar [30], is an incidence structure consisting of points, lines and planes, together with an incidence relation subject to the following conditions :

(i) If a point p is incident with a line L, and L is incident with a plane π, then p is incident with π.

(ii) Every two intersecting lines are coplanar.

(iii) Points and lines incident with a plane form a Bruck-net of order s+1 and degree t+1.

(iv) Planes are partitioned into b parallel classes such that :

(a) two planes from distinct classes intersect in a unique line,

(b) each point is in exactly one plane of each class.

(v) Each line is in at least one plane.

It is shown in [33] that a 3-net is a partial three-space. The following 3-net is due to R. Laskar and J.W. Freeman [31]. Let $PG(2,p^r)$ be a subplane of $PG(2,p^h)$, and let $PG(2,p^h)$ be embedded in $PG(3,p^h)$. Points of S are the points of $PG(3,p^h)-PG(2,p^h)$. Lines of S are the lines of $PG(3,p^h)$ which are not contained in $PG(2,p^h)$ and meet $PG(2,p^r)$. Planes of S are the planes of $PG(3,p^h)$ intersecting $PG(2,p^h)$ in a line of $PG(2,p^r)$. Incidence is the natural one. Then S is a 3-net with $s=p^h-1$, $t=p^r$, $b=p^{2r}+p^r+1$.

Another 3-net is constructed as follows [30]. Points are the ordered triples (i,j,k), $i,j,k\in\{1,\ldots,n\}=V$. Lines are the sets $L^3_{ij}=\{(i,j,k) \parallel k\in V\}$, $L^2_{ik}=\{(i,j,k) \parallel j\in V\}$ and $L^1_{jk}=\{(i,j,k) \parallel i\in V\}$. Planes are the sets $\pi^1_i=\{(i,j,k) \parallel j,k\in V\}$,

$\pi_j^2=\{(i,j,k) \parallel i,k\in V\}$ and $\pi_k^3=\{(i,j,k) \parallel i,j\in V\}$. Incidence is the natural one. Here we have $s=n-1$, $t=1$, $b=3$. The point graph of this 3-net is the cubic lattice graph.

A beautiful result of A.P. Sprague [42] tells us that there are no other (finite) 3-nets.

Finally we remark that 3-nets have been generalized to d-nets [25].

(2) <u>The classical partial three-spaces</u>. (a) The points, lines and planes of $PG(n,q)$, $n\geq 3$, form a partial three-space with parameters $s=t=q$, $\alpha=q+1$, $s^*=q^{n-2}+q^{n-3}+\ldots+q$, $\alpha^*=q+1$, $u=0$, $w=q^2+q+1$ if $n>3$ and w is not defined for $n=3$, u^* and w^* are not defined, $x=q^2+q+1$. We shall use the notation $PG(n,q)$ for this partial three-space.

(b) Let Q be a non-singular hyperquadric of index three of $PG(n,q)$, $n=5,6$ or 7. Then the points of Q together with the lines and planes of Q form a partial three-space $Q(n,q)$ with parameters [22]

$s=t=q$, $\alpha=q+1$, $s^*=1$, $\alpha^*=1$, $u=0$, $w=1$, $w^*=u^*=1$, $x=1$, when $n=5$;

$s=t=q$, $\alpha=q+1$, $s^*=q$, $\alpha^*=1$, $u=0$, $w=1$, $w^*=u^*=1$, $x=1$, when $n=6$;

$s=t=q$, $\alpha=q+1$, $s^*=q^2$, $\alpha^*=1$, $u=0$, $w=1$, $w^*=u^*=1$, $x=1$, when $n=7$.

(c) Let H be a non-singular hermitian variety of the projective space $PG(n,q^2)$, $n=5$ or 6. Then the points of H together with the lines and planes of H form a partial three-space $H(n,q^2)$ with parameters [22]

$s=t=q^2$, $\alpha=q^2+1$, $s^*=q$, $\alpha^*=1$, $u=0$, $w=1$, $w^*=u^*=1$, $x=1$, when $n=5$;

$s=t=q^2$, $\alpha=q^2+1$, $s^*=q^3$, $\alpha^*=1$, $u=0$, $w=1$, $w^*=u^*=1$, $x=1$, when $n=6$.

(d) The points of PG(5,q), together with the totally isotropic lines and planes with respect to a symplectic polarity of PG(5,q) form a partial three-space W(5,q) with parameters [22] $s=t=q$, $\alpha=q+1$, $s^*=q$, $\alpha^*=1$, $u=0$, $w=1$, $u^*=w^*=1$, $x=1$.

These partial three-spaces are due to R. Laskar and J.A. Thas [32].

(3) <u>The partial three-space T(n,q)</u>. The partial three-space T(n,q) is formed by the points not in a given $PG(n-3,q) \subset PG(n,q)$, $n \geq 3$, and by the lines and planes of PG(n,q) which have no point in common with PG(n-3,q). In fact the partial three-spaces T(n,q) are isomorphic to the duals of the Freeman-Laskar 3-nets.

<u>Remarks</u>. 1. Several other examples of partial three-spaces are known [32].

2. Let S be a classical partial three-space or a partial three-space T(n,q). Then the planes of S are planes of a projective space $PG(n,p^h)$, and the points and lines of S are all the points and lines of $PG(n,p^h)$ in these planes. The lines of the Freeman-Laskar 3-net are lines of an affine space $AG(3,p^h)$, and the points of this 3-net are all the points of $AG(3,p^h)$ on these lines.

<u>Literature</u>. For a survey on the subject we refer to R. Laskar and J. Dunbar [33], and to R. Laskar and J.A. Thas [32].

2. EMBEDDING IN FINITE PROJECTIVE SPACES

2.1. Introduction

In many of the examples in 1, the lines of the geometry are lines of a projective space PG(n,q), and the points of the geometry are all the points of PG(n,q) on these lines. We say that these geometries are embedded in a projective space. This part of the paper is concerned with the determination of the generalized quadrangles, partial geometries, semi partial geometries and partial three-spaces, which are embeddable in a finite projective space.

2.2. Generalized quadrangles in finite projective spaces

The following fundamental and beautiful result is due to F. Buekenhout and C. Lefèvre [9].

Theorem. If a point set of the projective space PG(n,q), together with a line set of PG(n,q) form a generalized quadrangle S, then S is of classical type (i.e. is one of Q(d,q), d=3,4 or 5, H(d,q), d=3 or 4, or W(3,q)).

Remark. In [37] , [38] ,[51] there are partial solutions of the same problem. I also remark that in [9] F. Buekenhout and C. Lefèvre were on their way to solve the problem in the infinite (commutative) case. Recently that question was completely solved by K.J.Dienst [24] .
Finally I notice that the reading of [9] gave me the idea to attack the problem for more general geometries.

2.3. Partial geometries in finite projective spaces

Theorem (F. De Clerck and J.A. Thas [21]). If S=(P,B,I) is a partial geometry with parameters s,t,α which is embeddable in a projective space PG(n, s), but not in a PG(n',s) with $n'<n$, then the following cases may occur :

(a) $\alpha=s+1$, and S is the design of points and lines of PG(n,s);

(b) $\alpha=1$, and S is a classical generalized quadrangle;

(c) $\alpha=t+1$, $n=2$, and S is a dual design in PG(2,s) (S=PG(2,s), S is the dual of AG(2,s), or the points not in S constitute a maximal arc of PG(2,s));

(d) $\alpha=s$ and $S=H_s^n$ ($n\geq 3$).

2.4. <u>Semi partial geometries in finite projective spaces</u>

<u>Theorem (I. Debroey and J.A. Thas [17])</u>. If S=(P,B,I) is a semi partial geometry with parameters s,t,α and μ, which is embedded in the projective space PG(3,s), then the following cases may occur :

(a) S is a dual design in a PG(2,s)$\subset$PG(3,s);

(b) S is a classical generalized quadrangle;

(c) S is the design of points and lines of PG(3,s);

(d) $S=H_s^3$;

(e) S is the semi partial geometry $\bar{W}(3,s)$ arising from a symplectic polarity of PG(3,s);

(f) S is a Desargues configuration in PG(3,2) (here $t=s=\alpha=2$, $\mu=4$).

<u>Remarks</u>. 1. This result has been used by J.W.P. Hirschfeld and J.A. Thas [29] in their classification of sets of type (1,n,q+1) in PG(d,q).

2. The difficulty in determining all semi partial geometries embeddable in PG(n,q), $n\geq 4$, is the following. Let S be a semi partial geometry in PG(n,q), $n\geq 4$, and let PG(3,q) be a subspace of PG(n,q) spanned by three concurrent lines of S (or by two concurrent lines and a point of S, or by

a line and two points of S, or by two non-concurrent lines of S). Then the structure S' induced by S on the space PG(3,q), is not a priori a semi partial geometry, and so we cannot apply the preceding theorem (in contrast with the partial geometries).

3. The only known semi partial geometry embeddable in PG(n,q), $n \geq 4$, is $\bar{W}(n,q)$ (here n is odd).

2.5. Partial three-spaces in finite projective spaces

If S is a partial three-space which is embedded in a projective space PG(n,q), then, to avoid confusion the points, lines and planes of S are called S-points, S-lines and S-planes respectively. A partial three-space is embedded in a PG(n,q) if the S-lines are lines of PG(n,q), if the S-points are all the points of PG(n,q) on these S-lines, and if the incidence relation is that of PG(n,q); the partial three-space is strong embedded in PG(n,q) if the S-planes are planes of PG(n,q), if the S-lines and S-points are all the lines and points of PG(n,q) in these S-planes, and if the incidence is that of PG(n,q). Evidently, first of all, we have to classify all partial three-spaces which are strong embeddable in a PG(n,q), before to attack the difficult general problem.

Theorem (J.A. Thas [44]). Let S be a partial three-space for which the S-planes are planes of PG(n,q), for which the S-lines are all the lines contained in the S-planes, for which the S-points are all the points in the S-planes, and for which the incidence relation is that of PG(n,q). We also suppose that the set of the S-points is not contained in a $PG(n',q) \subset PG(n,q)$, $n' < n$. Then the following cases can occur.

(a) S is a classical partial three-space;

(b) S is the partial three-space $T(n,q)$.

Remark. Essential for the proof of this theorem are the "Foundations of polar geometry" by F. Buekenhout and E. Shult [10], the classification of all partial geometries in finite projective spaces, and some results on sets of type (or class) $(1,n,q+1)$ in $PG(3,q)$ [29].

Main theorem (J.A. Thas [44]). Let S be a partial three-space for which the S-lines are lines of $PG(n,q)$, for which the S-points are all the points on the S-lines, for which the incidence relation is that of $PG(n,q)$, and for which the set of the S-points is not contained in a $PG(n',q) \subset PG(n,q)$, $n'<n$. Then one of the following cases occurs.

(a) S is a classical partial three-space (i.e. $PG(n,q)$, $Q(n,q)$ with $n\in\{5,6,7\}$, $H(n,q)$ with $n\in\{5,6\}$, or $W(5,q)$).

(b) S is the partial three-space $T(n,q)$.

(c) Let $\pi_1,\ldots,\pi_5$ be 5 planes of $PG(3,2)$ no four of which have a point in common. The S-lines are the intersections of the 10 pairs (π_i,π_j), $i\neq j$, the S-points are the 10 points on these S-lines, the S-planes are the planes $\pi_1,\ldots,\pi_5$, and incidence is that of $PG(3,2)$. The geometry of the S-points and S-lines incident with the S-plane π_i is the dual of a 2-(4,2,1) design. Finally, S is isomorphic to the partial three-space arising from the tetrahedral graph with characteristic 5 [33].

(d) The S-points and S-lines are the points and lines of the hyperbolic quadric Q in $PG(7,q)$, the S-planes are the three-spaces of one of the families of generating three-spaces of Q, and incidence is that of $PG(7,q)$.

Here the geometry of the S-points and S-lines incident with an S-plane is the design of points and lines of a $PG(3,q)$.

(e) S is isomorphic to the dual of one of the partial three-spaces $W(5,q)$, $Q(6,q)$, $Q(7,\sqrt{q})$, $H(5,q^2)$. Here the geometry of the S-points and S-lines incident with an S-plane is a classical generalized quadrangle.

(f) Consider the Segre-manifold [2] $V \subset PG(7,q)$ of a $PG(3,q)$ and a line (or, equivalently, consider a 3-regulus in $PG(7,q)$ [22]). The generating three-spaces are denoted by $PG^{(1)}(3,q),\ldots,PG^{(q+1)}(3,q)$. In $PG^{(1)}(3,q)$ we consider a hyperbolic quadric Q_1, with reguli $R_1 = \{L_1^1,\ldots,L_1^{q+1}\}$ and $U_1 = \{M_1^1,\ldots,M_1^{q+1}\}$. Suppose that $R_i = \{L_i^1,\ldots,L_i^{q+1}\}$ (resp. $U_i = \{M_i^1,\ldots,M_i^{q+1}\}$) is the regulus in $PG^{(i)}(3,q)$, $i=1,\ldots,q+1$, which corresponds to R_1 (resp. U_1), where notations are chosen in such a way that $V_j = \{L_1^j,\ldots,L_{q+1}^j\}$ (resp. $W_j = \{M_1^j,\ldots,M_{q+1}^j\}$), $j=1,\ldots$ $\ldots,q+1$, is a regulus. The complementary regulus of V_j (resp. W_j) is denoted by X_j (resp. Y_j). The S-points are the points on $R_1,\ldots,R_{q+1}$, the S-lines are the lines of R_i, U_i, X_i, $i=1,\ldots,q+1$, the S-planes are the hyperbolic quadrics with reguli R_i and U_i, the hyperbolic quadrics with reguli V_i and X_i, and the hyperbolic quadrics with reguli W_i and Y_i, $i=1,\ldots,q+1$. The incidence is that of $PG(7,q)$. Finally, S is isomorphic to the 3-net arising from the cubic lattice graph with $(q+1)^3$ vertices.

<u>Remarks</u>. 1. It is easy to check that cases (a), (b), (c), (d), (f) really occur. But also case (e) in the statement of the theorem really occurs. Indeed, we show there exists a partial three-space S which is

isomorphic to the dual of Q(6,q), for which the S-lines are lines of PG(7,q) and for which the S-points are all the points of PG(7,q) on these lines.

Consider the hyperbolic quadric Q^* in PG(7,q), and let PG(6,q) be a hyperplane for which $PG(6,q) \cap Q^* = Q$ is non-singular. Denote by R one of the families of generating three-spaces of Q^*. Any plane of Q is contained in exactly one element of R, and every element of R intersects PG(6,q) in a plane of Q. If L is a line of Q and if $\pi_1, \ldots, \pi_{q+1}$ are the q+1 planes of Q through L, then the elements of R containing these planes are exactly the q+1 elements of R containing L. By triality [53], with the planes of Q(6,q) correspond the points of Q^*, and with the lines of Q(6,q) correspond lines of Q^*. Hence the dual of Q(6,q) is isomorphic to a partial three-space S for which the S-lines are lines of Q^* and for which the S-points are all the points of Q^*.

Analogously one shows that the dual of $Q(7,\sqrt{q})$ is isomorphic to a partial three-space S for which the S-lines are lines of Q^* and for which the S-points are all the points of PG(7,q) on these lines (here the set of all S-points is not the complete point set of Q^*).

2. The proof of the main theorem shows that there are many links with the theory of sets of class (or type) (0,1,d,q+1) in PG(n,q) [29], [35], [43], with the theory of polar spaces [10], and with the near hexagons of E. Shult and A. Yanushka [41].

3. EMBEDDING IN FINITE AFFINE SPACES

3.1. Introduction

In several of the examples in 1, the lines of the geometry are lines of an affine space AG(n,q), and the points of the geometry are all the points of AG(n,q) on these lines. We say that these geometries are embedded in an affine space. This part of the paper is concerned with the determination of the generalized quadrangles, partial geometries and semi partial geometries, which are embeddable in a finite affine space.

3.2. Generalized quadrangles in finite affine spaces

The following complete classification of all generalized quadrangles embeddable in an AG(n,q) is in J.A. Thas [46].

Embedding in AG(2,s+1). If the generalized quadrangle S with parameters s,t is embedded in AG(2,s+1), then S is a net of order s+1 and degree 2.

Embedding in AG(3,s+1). Suppose that the generalized quadrangle S=(P,B,I) with parameters s,t is embedded in AG(3,s+1), and that P is not contained in a plane of AG(3,s+1). Then the following cases can occur.

(a) s=1, t=2 (trivial case);

(b) t=1 and the elements of S are the affine points and affine lines of an hyperbolic quadric of PG(3,s+1), the projective completion of AG(3,s+1), which is tangent to the plane at infinity of AG(3,s+1);

(c) s=2, t=2 (an embedding of the unique [52] generalized quadrangle with 15 points and 15 lines in AG(3,3));

(d) $S=T_2^*(O)$, i.e. P is the point set of AG(3,s+1), and B is the set of all lines of AG(3,s+1) whose points at infinity are the points of a complete oval O of the plane at infinity of AG(3,s+1);

(e) $S=W^*(3,s+1)$, i.e. P is the point set of AG(3,s+1) and $B=B_1 \cup B_2$, where B_1 is the set of all affine totally isotropic lines with respect to a symplectic polarity π of the projective completion PG(3,s+1) of AG(3,s+1) and where B_2 is the class of parallel lines defined by the pole (the image with respect to π) of the plane at infinity of AG(3,s+1).

Embedding in AG(4,s+1). Suppose that the generalized quadrangle S=(P,B,I) with parameters s,t is embedded in AG(4,s+1), and that P is not contained in an AG(3,s+1). Then the following cases can occur.

(a) s=1, $t \in \{2,3,4,5,6,7\}$ (trivial cases);

(b) s=t=2 (an embedding of the unique generalized quadrangle with 15 points and 15 lines in AG(4,3));

(c) s=t=3 and S is isomorphic to the generalized quadrangle Q(4,3) arising from a non-singular hyperquadric of PG(4,3);

(d) s=2, t=4 (an embedding of the unique [52] generalized quadrangle with 27 points and 45 lines in AG(4,3)).

Embedding in AG(n,s+1), $n>4$. Suppose that the generalized quadrangle S=(P,B,I) with parameters s,t is embedded in AG(n,s+1), $n \geq 5$, and that P is not contained in an AG(n-1,s+1). Then the following cases can occur.

(a) s=1 and $t \in \{[n/2], \ldots, 2^{n-1}-1\}$ (trivial case);

(b) s=2, t=4, n=5 (an embedding of the unique generalized quadrangle with 27 points and 45 lines in AG(5,3)).

Description of the five sporadic cases

1. s=t=2, n=3. Let ω be a plane of AG(3,3), and let

$\{L_0, L_1, L_2\}$ and $\{M_x, M_y, M_z\}$ be two classes of parallel lines of ω. Suppose that $\{x_i\} = M_x \cap L_i$, $\{y_i\} = M_y \cap L_i$ and $\{z_i\} = M_z \cap L_i$. Further, let N_x, N_y, N_z be three lines containing respectively x_0, y_0, z_0, for which $N_x \notin \{M_x, L_0\}$, $N_y \notin \{M_y, L_0\}$, $N_z \notin \{M_z, L_0\}$, for which the planes $N_x M_x$, $N_y M_y$, $N_z M_z$ are parallel, and for which $\omega, L_0 N_x, L_0 N_y, L_0 N_z$ are distinct. The points of N_x are x_0, x_3, x_4, the points of N_y are y_0, y_3, y_4, and the points of N_z are z_0, z_3, z_4, where notations are chosen in such a way that x_3, y_3, z_3, resp. x_4, y_4, z_4 are collinear. The points of the generalized quadrangle are $x_0, \ldots, x_4$, $y_0, \ldots, y_4, z_0, \ldots, z_4$, and the lines are $L_0, L_1, L_2, M_x, M_y, M_z$, N_x, N_y, N_z, $x_3 y_4$, $x_4 y_3$, $x_3 z_4$, $x_4 z_3$, $y_3 z_4$, $y_4 z_3$.

2. <u>$s=t=2$, $n=4$</u>. Let $PG(3,3)$ be the hyperplane at infinity of $AG(4,3)$, let ω_∞ be a plane of $PG(3,3)$, and let 1 be a point of $PG(3,3) - \omega_\infty$. In ω_∞ we choose points m_{01}, m_{02}, $m_{11}, m_{12}, m_{21}, m_{22}$, in such a way that m_{01}, m_{21}, m_{11} are collinear, that m_{11}, m_{02}, m_{22} are collinear, that m_{21}, m_{02}, m_{12} are collinear, and that m_{01}, m_{22}, m_{12} are collinear. L is an affine line containing 1, and the affine points of L are denoted by p_0, p_1, p_2. The points of the generalized quadrangle are the affine points of the lines $p_0 m_{01}$, $p_0 m_{02}, p_1 m_{11}, p_1 m_{12}$, $p_2 m_{21}, p_2 m_{22}$. The lines of the quadrangle are the affine lines of the hyperbolic quadric containing $p_0 m_{01}, p_1 m_{11}, p_2 m_{21}$, the affine lines of the hyperbolic quadric containing $p_0 m_{02}, p_1 m_{11}, p_2 m_{22}$, the affine lines of the hyperbolic quadric containing $p_0 m_{02}, p_1 m_{12}, p_2 m_{21}$, and the affine lines of the hyperbolic quadric containing $p_0 m_{01}, p_1 m_{12}, p_2 m_{22}$.

3. <u>$s=t=3$, $n=4$</u>. Let $PG(3,4)$ be the hyperplane at infinity of $AG(4,4)$, let ω_∞ be a plane of $PG(3,4)$, let H be a hermitian curve (or a unital) of ω_∞, and let 1 be a point

of $PG(3,4)-\omega_\infty$. In ω_∞ there are exactly 4 triangles $m_{01}m_{02}m_{03}, m_{11}m_{12}m_{13}, m_{21}m_{22}m_{23}, m_{31}m_{32}m_{33}$ whose vertices are exterior points of H and whose sides are secants of H. Any line $m_{0a}m_{1b}$, $a,b\in\{1,2,3\}$, contains exactly one vertex m_{2c} of $m_{21}m_{22}m_{23}$, and one vertex m_{3d} of $m_{31}m_{32}m_{33}$ ($m_{0a}m_{1b}$ is a tangent line of H). We remark that the cross-ratio $(m_{0a}m_{1b}m_{2c}m_{3d})$ is independent of the choice of $a,b\in\{1,2,3\}$. Let L be an affine line through 1, and let p_0, p_1, p_2, p_3 be the affine points of L, where notations are chosen in such a way that $(p_0p_1p_2p_3)=(m_{0a}m_{1b}m_{2c}m_{3d})$. The points of the generalized quadrangle are the 40 affine points of the lines $p_0m_{01}, p_0m_{02}, p_0m_{03}, p_1m_{11}, p_1m_{12}, p_1m_{13}, p_2m_{21}, p_2m_{22}, p_2m_{23}, p_3m_{31}, p_3m_{32}, p_3m_{33}$. The lines of the quadrangle are the affine lines of the hyperbolic quadric containing $p_0m_{0a}, p_1m_{1b}, p_2m_{2c}, p_3m_{3d}$, $a,b=1,2,3$.

4. <u>$s=2$, $t=4$, $n=4$</u>. Let $PG(3,3)$ be the hyperplane at infinity of $AG(4,3)$, let ω_∞ be a plane of $PG(3,3)$ and let 1 be a point of $PG(3,3)-\omega_\infty$. In ω_∞ we choose points $m, n_x, n_y, n_z, n'_x, n'_y, n'_z, n''_x, n''_y, n''_z$ in such a way that m, n_x, n_y, n_z are collinear, that m, n'_x, n'_y, n'_z are collinear, that m, n''_x, n''_y, n''_z are collinear, and that n_a, n'_b, n''_c, with $\{a,b,c\}=\{x,y,z\}$, are collinear. Let L be an affine line through 1, and let x,y,z be the affine points of L. ω is the plane defined by L and m. The points of the generalized quadrangle are the 27 affine points of the lines $xm, xn_x, xn'_x, xn''_x, ym, yn_y, yn'_y, yn''_y, zm, zn_z, zn'_z, zn''_z$. The 45 lines of the quadrangle are the affine lines of ω with points at infinity 1 and m, the affine lines of the hyperbolic quadric containing am, bn_b, cn_c (resp. am, bn'_b, cn'_c, resp. am, bn''_b, cn''_c), with $\{a,b,c\}=\{x,y,z\}$, and the affine lines of the hyperbolic quadric containing an_a, bn'_b, cn''_c, with $\{a,b,c\}=\{x,y,z\}$.

5. <u>s=2, t=4, n=5</u>. Let PG(4,3) be the hyperplane at infinity of AG(5,3), let H_∞ be a hyperplane of PG(4,3) and let 1 be a point of PG(4,3)-H_∞. In H_∞ we choose points $m_x, m_y, m_z, n_x, n_y, n_z, n'_x, n'_y, n'_z, n''_x, n''_y, n''_z$ in such a way that m_x, m_y, m_z are collinear, that $m_x, m_y, m_z, n_x, n_y, n_z$ are in a plane ω_∞, that $m_x, m_y, m_z, n'_x, n'_y, n'_z$ are in a plane ω'_∞, that $m_x, m_y, m_z, n''_x, n''_y, n''_z$ are in a plane ω''_∞, that m_a, n_b, n_c (resp. m_a, n'_b, n'_c, resp. m_a, n''_b, n''_c), with $\{a,b,c\}=\{x,y,z\}$, are collinear, and that n_a, n'_b, n''_c, with $\{a,b,c\}=\{x,y,z\}$, are collinear. Let L be an affine line through 1, and let x,y,z be the affine points of L. The points of the generalized quadrangle are the 27 affine points of the lines xm_x, ym_y, zm_z, xn_x, $yn_y, zn_z, xn'_x, yn'_y, zn'_z$, xn''_x, yn''_y, zn''_z. The 45 lines of the quadrangle are the affine lines of the hyperbolic quadric containing xm_x, ym_y, zm_z, the affine lines of the hyperbolic quadric containing am_a, bn_b, cn_c (resp. am_a, bn'_b, cn'_c, resp. am_a, bn''_b, cn''_c), with $\{a,b,c\}=\{x,y,z\}$, and the affine lines of the hyperbolic quadric containing an_a, bn'_b, cn''_c, with $\{a,b,c\}=\{x,y,z\}$.

<u>Remark</u>. The embedding in AG(3,s+1) is also considered in [5].

3.3. <u>Partial geometries in finite affine spaces</u>

The following complete classification of all partial geometries embeddable in AG(n,q) is in J.A. Thas [46] .

<u>Embedding in AG(2,s+1)</u>. If the partial geometry S=(P,B,I) with parameters s,t,α is embedded in AG(2,s+1), then S is a net of order s+1 and degree t+1, or B∪{line at infinity of AG(2,s+1)} is a complete oval of the dual projective plane of PG(2,s+1), where PG(2,s+1) is the projective

completion of $AG(2,s+1)$ (here $s=2^h-1$, $t=1$, $\alpha=2$).

Embedding in $AG(3,s+1)$. Suppose that the partial geometry $S=(P,B,I)$ with parameters s,t,α, where $\alpha>1$, is embedded in $AG(3,s+1)$, and that P is not contained in a plane of $AG(3,s+1)$. Then the following cases can occur.

(a) $s=1$, $\alpha=2$, $t\in\{2,3,4,5\}$ (S is a 2-$(t+2,2,1)$ design in $AG(3,2)$);

(b) P is the point set of $AG(3,s+1)$, and B is the set of all lines of $AG(3,s+1)$ whose points at infinity are the points of a (maximal) $\{(s+1)d-(s+1)+d;d\}$-arc K, $2<d\leq s+2$, of the plane at infinity of $AG(3,s+1)$ (here $S=T_2^*(K)$).

Embedding in $AG(n,s+1)$, $n\geq 4$. Suppose that the partial geometry $S=(P,B,I)$ with parameters s,t,α, with $\alpha>1$, is embedded in $AG(n,s+1)$, where $n\geq 4$, and that P is not contained in an $AG(n',s+1)$, with $n'<n$. Then the following cases can occur.

(a) $s=1$, $\alpha=2$, $t\in\{n-1,n,\dots,2^n\}$ and then S is a 2-$(t+2,2,1)$ design (P is an arbitrary point set of $AG(n,2)$ which is not contained in an $AG(n',2)$, $n'<n$);

(b) S is the design of points and lines of $AG(n,s+1)$;

(c) P is the point set of $AG(n,s+1)$, and B is the set of all lines of $AG(n,s+1)$ whose points at infinity constitute the complement of a hyperplane $PG(n-2,s+1)$ of the space at infinity of $AG(n,s+1)$ (here $t=(s+1)^{n-1}-1$ and $\alpha=s$).

3.4. Semi partial geometries in finite affine spaces

Theorem (I. Debroey and J.A. Thas [19]). If $S=(P,B,I)$ is a semi partial geometry with parameters s,t,α and μ, which is embeddable in the projective spaces $AG(2,s+1)$ or

AG(3,s+1), then the following cases may occur.

(a) S is a partial geometry (and then S is known by 3.3);

(b) S is a pentagon in AG(3,2) (trivial case);

(c) $S=T_2^*(B')$, i.e. the semi partial geometry arising from a Baer subplane B' of PG(2,s+1);

(d) $S=T_2^*(U)$, i.e. the semi partial geometry arising from a unital U of PG(2,s+1).

Remarks. 1. The proof of the theorem is long and tricky. Probably the determination of all semi partial geometries embeddable in AG(n,q), $n\geq 4$, is hopeless.

2. Now we give a description of all known semi partial geometries, which are not partial geometries and which are embeddable in an AG(n,s+1), $n\geq 4$ and $s>1$. (Evidently every semi partial geometry with s=1 is embeddable in an AG(n,2).)

(a) The following construction is in [29]. Let Q be a non-singular elliptic hyperquadric of PG(5,s+1), $s+1=2^h$. Consider a point $p\notin Q$ and a hyperplane H of PG(5,s+1) which does not contain p. Call Q' the projection of Q from p onto H, and call H' the set of the intersections of H with the tangent lines of Q containing p. Since $s+1=2^h$ the set H' is an hyperplane of H. The point set of the semi partial geometry is Q'-H'=P, the line set of the geometry is the set B of all affine lines of AG(4,s+1)=H-H' contained in Q'-H', and the incidence relation I is the natural one. The semi partial geometry (P,B,I), which is embedded in an AG(4,s+1), has parameters $s,t=(s+1)^2$, $\alpha=2$ and $\mu=2s(s+1)$.

(b) In [11] P.J. Cameron gives the following construction of partial quadrangles (i.e. semi partial geometries with $\alpha=1$). Let K be an r-cap of PG(n-1,s+1) (i.e. a set of r points no three of which are collinear) with the property that any point of PG(n-1,s+1)-K lies on θ tangents to K. Suppose that PG(n-1,s+1) is the hyperplane at infinity of AG(n,s+1). P is the point set of AG(n,s+1), B is the set of all lines of AG(n,s+1) whose points at infinity are the points of K, and I is the natural incidence relation. Then (P,B,I) is a partial quadrangle with parameters s, $t=r-1$, $\mu=r-\theta$.

For $n\geq 4$ the following examples are known [28]:

$n=4$, K is an ovoid O of PG(3,s+1), $r=(s+1)^2+1$, $\theta=(s+1)+1$ (here $S=T_3^*(O)$);

$n=5$, $s=2$, $r=11$, $\theta=9$ (see also [14], [4], [45]);

$n=6$, $s=2$, $r=56$, $\theta=36$.

3.5. Partial three-spaces in finite affine spaces

Some examples are known [32], but there are no classification theorems.

State University of Ghent

REFERENCES

[1] R.Ahrens and G. Szekeres, On a combinatorial generalization of 27 lines associated with a cubic surface, J.Austral. Math. Soc. 10 (1969), 485-492.

[2] H.F. Baker, Principles of geometry, Cambridge at the University Press, 1933.

[3] A. Barlotti, Sui {k;n}-archi di un piano lineare finito, Boll. Un. Mat. Ital. (3) 11 (1956), 553-556.

[4] E.R. Berlekamp, J.H. van Lint, and J.J. Seidel, A strongly regular graph derived from the perfect ternary Golay code, in A Survey of Combinatorial Theory, ed. by J.N. Srivastava et al. (1973), 25-30.

[5] A. Bichara, Caratterizzazione dei sistemi rigati immersi in $A_{3,q}$, Riv. Mat. Univ. Parma, to appear.

[6] A Bichara, F. Mazzocca, and C. Somma, Sulla classificazione dei sistemi rigati immersi in AG(3,q), to appear.

[7] R.C. Bose, Strongly regular graphs, partial geometries, and partially balanced designs, Pacific J. Math. 13 (1963), 389-419.

[8] R.C. Bose, Graphs and designs, C.I.M.E., II ciclo, Bressanone 1972, 1-104.

[9] F. Buekenhout and C. Lefèvre, Generalized quadrangles in projective spaces, Arch. Math. 25 (1974), 540-552.

[10] F. Buekenhout and E. Shult, Foundations of polar geometry, Geometriae Dedicata 3 (1974), 155-170.

[11] P.J. Cameron, Partial quadrangles, Quart. Journ. Math. Oxford (3) 25 (1974), 1-13.

[12] P.J. Cameron, J.M. Goethals, and J.J.Seidel, Strongly regular graphs having strongly regular subconstituents, J. Algebra, to appear.

[13] A Cossu, Su alcune proprietà dei {k;n}-archi di un piano proiettivo sopra un corpo finito, Rend. Mat. e Appl. 20 (1961), 271-277.

[14] H.S.M. Coxeter, Twelve points in PG(5,3) with 95040 self-transformations, Proc. Roy. Soc. London (A) 247 (1958), 279-293.

[15] I. Debroey, Semi partial geometries satisfying the diagonal axiom, J. of Geometry, to appear.

[16] I. Debroey, Semi partiële meetkunden, Ph. D. Dissertation, Rijksuniversiteit te Gent (1978).

[17] I. Debroey and J.A. Thas, Semi partial geometries in PG(2,q) and PG(3,q), Rend. Accad. Naz. Lincei, to appear.

[18] I. Debroey and J.A. Thas, On semi partial geometries, J. Comb. Theory, to appear.

[19] I. Debroey and J.A. Thas, Semi partial geometries in AG(2,q) and AG(3,q), Simon Stevin 51 (1977-78), 195-209.

[20] F. De Clerck, Partiële meetkunden, Ph. D. Dissertation, Rijksuniversiteit te Gent (1978).

[21] F. De Clerck and J.A. Thas, Partial geometries in finite projective spaces, Arch. Math. 30 (1978), 537-540.

[22] P. Dembowski, Finite geometries, Springer-Verlag, 1968.

[23] R.H.F. Denniston, Some maximal arcs in finite projective planes, J. Comb. Theory 6 (1969), 317-319.

[24] K.J.Dienst, Verallgemeinerte Vierecke in Pappusschen Projektiven Räumen, Technische Hochschule Darmstadt, Preprint-Nr.398 (1978).

[25] J. Dunbar and R. Laskar, Finite nets of dimension d, Discrete Math. 22 (1978), 1-24.

[26] M. Hall, Jr., Affine generalized quadrilaterals, in Studies in Pure Mathematics, ed. by L. Mirsky, Academic Press (1971), 113-116.

[27] D.G. Higman, Partial geometries, generalized quadrangles, and strongly regular graphs, in Atti Convegno di Geometria e sue Applicazioni, Perugia, 1971.

[28] R. Hill, Ph. D. Dissertation, University of Warwick (1971).

[29] J.W.P. Hirschfeld and J.A. Thas, Sets of type (1,n,q+1) in PG(d,q), Proc. London Math. Soc., to appear.

[30] R. Laskar, Finite nets of dimension three I, J. Algebra 32 (1974), 8-25.

[31] R. Laskar and J.W. Freeman, Further results on 3-nets, AMS Notices, Jan. 1977, A-35.

[32] R. Laskar and J.A. Thas, On some generalizations of partial geometry, in Graph Theory and Related Topics, ed.by J.A. Bondy and U.S.R. Murty, to appear.

[33] R. Laskar and J. Dunbar, Partial geometry of dimension three, J. Comb. Theory A 24 (1978), 187-201.

[34] C. Lefèvre, Semi-quadriques en tant que sous-ensembles des espaces projectifs, Bull. Soc. Math. de Belg. 29 (1977), 175-183.

[35] C. Lefèvre, Semi-quadriques et sous-ensembles des espaces projectifs, Ph. D. Dissertation, Université Libre de Bruxelles (1976).

[36] R. Nowakowski, private communication (1978).

[37] D. Olanda, Sistemi rigati immersi in uno spazio proiettivo, Ist. Mat. Univ. Napoli, Rel.n.26 (1973), 1-21.

[38] D. Olanda, Sistemi rigati immersi in uno spazio proiettivo, Rend. Accad. Naz. Lincei 62 (1977), 489-499.

[39] S.E. Payne, Finite generalized quadrangles : a survey, Proc. Int. Conf. Proj. Planes, Wash. State Univ. Press (1973), 219-261.

[40] S.E. Payne and J.A. Thas, Generalized quadrangles with symmetry, Simon Stevin 49 (1975-1976), 3-32 and 81-103.

[41] E. Shult and A. Yanushka, Near n-gons and equiangular line systems, to appear.

[42] A.P. Sprague, A characterization of 3-nets, J. Comb. Theory, to appear.

[43] G. Tallini, Problemi e resultati sulle geometrie di Galois, Ist. Mat. Univ. Napoli, Rel.n.30 (1973), 1-30.

[44] J.A. Thas, Partial three-spaces in finite projective spaces, to appear.

[45] J.A. Thas, Ovoids and spreads of finite classical polar spaces, Geometriae Dedicata, to appear.

[46] J.A. Thas, Partial geometries in finite affine spaces, Math. Z. 158 (1978), 1-13.

[47] J.A. Thas, Some results concerning $\{(q+1)(n-1);n\}$-arcs and $\{(q+1)(n-1)+1;n\}$-arcs in finite projective planes of order q, J. Comb. Theory 19 (1975), 228-232.

[48] J.A. Thas, Construction of maximal arcs and partial geometries, Geometriae Dedicata 3 (1974), 61-64.

[49] J.A. Thas, Combinatorics of partial geometries and generalized quadrangles, in Higher Combinatorics, ed. by M. Aigner (1977), 183-199.

[50] J.A. Thas and F. De Clerck, Partial geometries satisfying the axiom of Pasch, Simon Stevin 51 (1977), 123-137.

[51] J.A. Thas and P. De Winne, Generalized quadrangles in finite projective spaces, J. of Geometry 10 (1977), 126-137.

[52] J.A. Thas and S.E. Payne, Classical finite generalized quadrangles : a combinatorial study, Ars Combinatoria 2 (1976), 57-110.

[53] J. Tits, Sur la trialité et certains groupes qui s'en déduisent, Publ. Math. I.H.E.S., Paris 2 (1959), 14-60.

[54] J. Tits, Buildings and BN-pairs of spherical type, Lect. Notes Math. 386, Springer-Verlag 1974.

[55] W.D. Wallis, Configurations arising from maximal arcs, J. Comb. Theory (A) 15 (1973), 115-119.

8 · Long cycles in digraphs with constraints on the degrees

CARSTEN THOMASSEN

INTRODUCTION

There is an extensive literature on long cycles, in particular Hamiltonian cycles, in undirected graphs. Since an undirected graph may be thought of as a symmetric digraph it seems natural to generalize (some of) these results to digraphs. However, cycles in digraphs are more difficult to deal with than cycles in undirected graphs and there are relatively few results in this area. In the present paper we review the results on long cycles, in particular Hamiltonian cycles, in digraphs with constraints on the degrees, and we present a number of unsolved problems.

TERMINOLOGY

We use standard terminology with a few modifications explained below. A *digraph* (directed graph) consists of a finite set of *vertices* and a set of ordered pairs xy of vertices called *edges*. If the edge xy is present, we say that x *dominates* y. The *outdegree* $d^+(x)$ of x is the number of vertices dominated by x, the *indegree* $d^-(x)$ is the number of vertices dominating x, and the degree $d(x)$ of x is defined as $d^+(x) + d^-(x)$.

A digraph D is k-regular if each vertex has degree k and D is *k-diregular* if each vertex has indegree k and outdegree k. The *order* of a digraph is the number of vertices in it. When we speak of paths or cycles in digraphs, we always mean directed paths or cycles. A *k-cycle* is a cycle of length k. A digraph of order n is *pancyclic* if it contains a k-cycle for each $k = 2,3,\ldots,n$. A digraph with no 2-cycles is an *oriented graph* and an oriented graph with no two non-adjacent vertices is a *tournament*. A digraph is *strong* if, for any two vertices x,y, the digraph has a path from x to y and a path from y to x. A maximal strong subdigraph of a digraph D is called a *component* of D. A digraph is *k-connected* if the removal of fewer than k vertices always leaves a strong digraph. If G is an undirected graph, then G^* denotes the symmetric digraph associated with G.

LONG CYCLES IN DIGRAPHS

The first results on Hamiltonian cycles in digraphs are those of Camion and Ghouila-Houri.

THEOREM 1 (Camion [7]) A tournament is Hamiltonian if and only if it is strong.

THEOREM 2 (Ghouila-Houri [11]) A strong digraph of order n and minimum degree at least n is always Hamiltonian.

Ghouila-Houri's theorem implies the well-known theorem of Dirac [9] that every undirected graph of order n and minimum degree at least $n/2$ is always Hamiltonian. Ore [19] proved that the same conclusion holds if we only assume that the sum of degrees for any two non-adjacent vertices is at least n. This was generalized to digraphs by Woodall.

THEOREM 3 (Woodall [29]) If a digraph D of order n has the property that for any two vertices x,y, either x dominates y or

$$d^+(x) + d^-(y) \geq n \quad ,$$

then D is Hamiltonian.

A common generalization of Theorems 1, 2 and 3 was obtained by Meyniel.

THEOREM 4 (Meyniel [15]) A strong digraph D of order n is Hamiltonian if for any two non-adjacent vertices x and y we have

$$d(x) + d(y) \geq 2n-1 \quad .$$

A short proof of Meyniel's theorem was found by Overbeck-Larish [21] and by modifying her proof, Bondy and the author described an even simpler proof [6].

Theorems 2, 3 and 4 are the best possible in the sense that they become false if the condition on the degrees is relaxed by one. This can be demonstrated by the complete bipartite digraphs with the property that the difference between the cardinality of the colour-classes is one. In [26] it is pointed out that there even exists a strong non-Hamiltonian digraph of order n with only one pair of non-adjacent vertices each of degree $n-1$ such that all other vertices have degree at least $(3n-5)/2$. To see this we take two complete graphs of order k and select a vertex z and z' , respectively, in each of these. Then we add two new vertices x,y and let each of these dominate each vertex of the complete digraph containing z and be dominated by each vertex of the other complete digraph. Then we add all edges from the complete digraph containing z' to the complete digraph containing z and finally we identify z' and z . Then the resulting digraph has the desired properties with $n = 2k+1$.

There are various generalizations of Ore's theorem for undirected graphs. The first of these is Pósa's theorem [23]. It seems difficult to generalize Pósa's theorem to digraphs. Nash-Williams made the following conjecture:

CONJECTURE 1 (Nash-Williams [18]) A digraph D of order n is Hamiltonian if its vertices can be labelled $x_1, x_2, \ldots, x_n$ and $y_1, y_2, \ldots, y_n$ such that

$$d^+(x_i) \geq i+1 \quad \text{for} \quad 1 \leq i \leq (n-2)/2$$

$$d^+(x_i) \geq n/2 \quad \text{for} \quad n/2 \leq i \leq n$$

and

$$d^+(x_{(n-1)/2}) \geq (n-1)/2 \quad \text{if} \quad n \text{ is odd}$$

and also

$$d^-(y_i) \geq i+1 \quad \text{for} \quad 1 \leq i \leq (n-2)/2$$

$$d^-(y_i) \geq n/2 \quad \text{for} \quad n/2 \leq i \leq n$$

and

$$d^-(y_{(n-1)/2}) \geq (n-1)/2 \quad \text{if} \quad n \text{ is odd} .$$

Nash-Williams [18] also raised the question of finding a common generalization of Theorems 1 and 2 by characterizing all non-Hamiltonian strong digraphs of order n and minimum degree $n-1$. By Camion's theorem, none of these are tournaments. Bondy [4] conjectured that also none of these are $(n-1)/2$-diregular except D_5 and D_7 of Figure 1. Bondy [4] also made the stronger conjectures that every strong $(n-1)$-regular digraph (except D_5 and D_7) is Hamiltonian and that every k-diregular digraph of order $2k+1$ (except

D_5 and D_7) can be decomposed into Hamiltonian cycles.

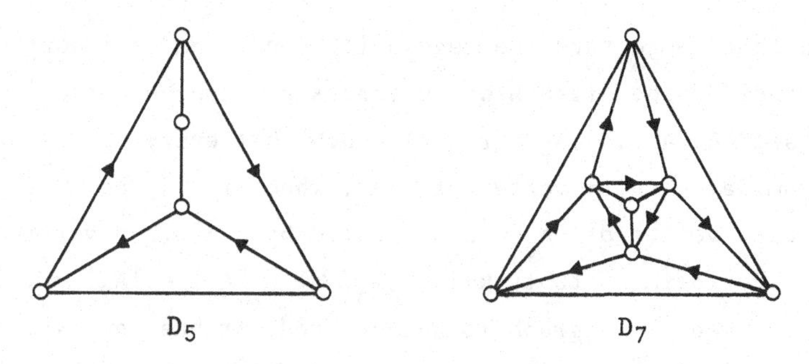

Figure 1: Two k-diregular non-Hamiltonian graphs of order 2k+1

The last of these conjectures would imply Kelly's conjecture:

CONJECTURE 2 (Kelly, see Moon [16]) Every diregular tournament can be decomposed into Hamiltonian cycles.

The above-mentioned conjecture of Bondy on Hamiltonian cycles in (n-1)-regular digraphs is not true. In [26] a variety of counter-examples are described. Consider for example the digraph with only one pair of non-adjacent vertices described in connection with Meyniel's theorem. It is easy to find a spanning strong (n-1)-regular subdigraph of this digraph. These examples and other examples in [26] indicate that it is not easy to describe the strong non-Hamiltonian digraphs of order n and minimum degree $n-1$. Among these, all the regular digraphs that we know of have connectivity one so the following conjecture may be true.

CONJECTURE 3 Every 2-connected (n-1)-regular digraph of order n, except D_5 and D_7, is Hamiltonian.

In this conjecture the regularity condition is important. For consider the digraph whose vertex set can be partitioned into set $A_o, A_1, \ldots, A_r$ $(r \geq 2)$ such that every vertex of A_i dominates every vertex of A_j, when $i < j$ and furthermore each vertex of A_o is dominated by all other vertices. Assume furthermore that $k+1 = |A_o| + 1 = |A_1| = |A_2| = \ldots = |A_r|$. Then the digraph is k-connected, it has $n = k+r(k+1)$ vertices and minimum degree $n-1$. However, any longest cycle of the digraph misses exactly one vertex from each A_i, $1 \leq i \leq r$, i.e. for any longest cycle, the vertices not in this cycle induce a transitive tournament of order r. In [26] it is shown that this situation is typical for 2-connected digraphs of order n and minimum degree $n-1$.

THEOREM 5 ([26]) If D is a strong digraph of order n and minimum degree $n-1$ and S is any longest cycle of D, then every vertex of $D-V(S)$ has degree $n-1$, any two vertices of $D-V(S)$ are adjacent, and every component of $D-V(S)$ is a complete digraph. Moreover, if D is 2-connected, then S can be chosen such that $D-S$ is a transitive tournament.

It is shown in [26] that a component of $D-S$ may have any order s, $s < n/2$.

We have already described many strong non-Hamiltonian digraphs of order $n = 2k+1$ and minimum degree $n-1$. If we impose the additional condition that all indegrees and outdegrees are greater than or equal to k, then the class of digraphs we get is considerably smaller.

THEOREM 6 ([26]) Let D be a digraph of order $n = 2k+1$ and minimum indegree and outdegree $\geq k$. Then D is Hamiltonian unless D has a set of $k+1$ mutually nonadjacend vertices (which then dominate and are dominated by all the k remaining vertices) or else D is isomorphic to D_5 or D_7 of Figure 1.

Since D_5 and D_7 are the only diregular digraphs in Theorem 6, the above-mentioned conjecture of Bondy on Hamiltonian cycles in diregular digraphs follows.

COROLLARY 1 ([26]) If D is a k-diregular digraph of order $n = 2k+1$, then D is Hamiltonian unless D is isomorphic to D_5 or D_7 of Figure 1.

However, Bondy's conjecture on Hamiltonian decomposition of diregular digraphs is not true. Tillson [28] has proved that the complete digraph K_{2m} can be decomposed into Hamiltonian cycles if and only if $m \neq 2,3$. Using the fact that K_4^* and K_6^* have no Hamiltonian decomposition we get the counter-examples to the conjecture of Bondy shown in Figure 2(b) and (c).

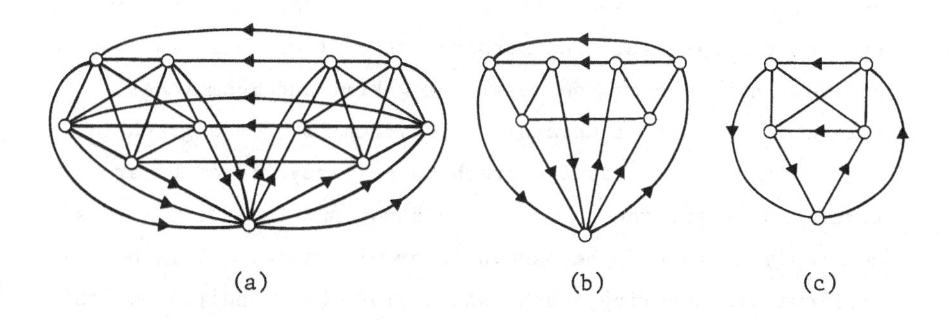

(a) (b) (c)

Figure 2: Three diregular digraphs with no Hamiltonian decomposition

We do not know of an infinite family of counter-examples to the conjecture, so maybe it becomes true for k sufficiently large. If one could specify the exceptional digraphs for the conjecture, this would settle Kelly's conjecture.

Another way of attacking Kelly's conjecture is suggested in [26].

CONJECTURE 4 ([26]) Every oriented graph of order n and minimum indegree and outdegree $\geq n/3$ is Hamiltonian.

If true, this conjecture at least would show that a diregular tournament has many edge-disjoint Hamiltonian cycles. The degree condition in the conjecture cannot be relaxed. For consider the digraph whose vertex set can be partitioned into sets A_1, A_2, A_3 such that a vertex x in A_i dominates a vertex y in A_j if and only if $j \equiv i+1 \pmod{3}$. If $|A_1|, |A_2|, |A_j|$ are almost equal (but not equal) this digraph is non-Hamiltonian and the minimum indegree and outdegree is almost $n/3$.

Bondy observed that conditions implying an undirected graph to be Hamiltonian often imply the graph to be pancyclic or to have a very special structure. Specifically, he proved that Ore's condition implies a graph to be pancyclic or bipartite. The strongest result of this kind for undirected graphs is the result of Hakimi and Schmeichel [12] who proved that Chvátal's condition [8] for a graph to be Hamiltonian implies the graph to be pancyclic or bipartite. (However, the condition of Bondy and Chvátal [5] does not imply a graph to be pancyclic or bipartite. It is not difficult to describe graphs satisfying this condition which are not bipartite and which have no 3-cycles.) The abovementioned result of Bondy was generalized to digraphs by the author.

THEOREM 7 ([24]) Let D be a strong digraph of order n such that for any two non-adjacent vertices x and y of D we have

$$d(x) + d(y) \geq 2n \quad .$$

Then D is pancyclic or D is a tournament (in which case D contains cycles of all lengths except 2) or else n is even and D is isomorphic to $K^*_{n/2,n/2}$.

This theorem extends Theorems 1, 2 and 3 (the fact that Ghouila-Houri's condition implies a digraph to be pancyclic or isomorphic to $K^*_{n/2,n/2}$ was first established by Häggkvist and the author [13]) and a result of Overbeck-Larish [22]. However, it does not include Meyniel's theorem and it becomes false, if we replace the degree condition by Meyniel's condition. To see this we consider the digraph $D_{n,k}$ with vertex set $\{x_1,x_2,\ldots,x_n\}$ and edge set $\{x_ix_j \mid i > j$ or $i = j-1\} \setminus \{x_ix_{i-k+1} \mid k \leq i \leq n\}$. This digraph has no cycle of length k and if $k > (n+1)/2$ then $D_{n,k}$ satisfies Meyniel's condition and it has only two pairs of non-adjacent vertices such that the inequality in Meyniel's theorem is an equality. Also when $k > (n+1)/2$, $D_{n,k}$ has proper subdigraphs satisfying Meyniel's condition.

The digraph $D_{n,k}$ above has some other interesting properties. Bondy [3] proved that every Hamiltonian undirected graph with n vertices and $n^2/4$ or more edges is pancyclic unless n is even and the graph is isomorphic to $K_{n/2,n/2}$, and he conjectured that the same is true for digraphs (when we replace $n^2/4$ by $n^2/2$ and $K_{n/2,n/2}$ by $K^*_{n/2,n/2}$). However, $D_{n,n-1}$ is Hamiltonian but not pancyclic and it has $(n(n+1)-6)/2$ edges. On the other hand we have

THEOREM 8 (Häggkvist and Thomassen [13]) A Hamiltonian

digraph with n vertices and $(n(n+1)-2)/2$ or more edges is pancyclic.

If $k > n$, then $D_{n,k}$ has $(n(n+1)-2)/2$ edges and contains exactly one Hamiltonian cycle.

THEOREM 9 (Müller and Pelant [17]) A Hamiltonian digraph with n vertices and at least $n(n+1)/2$ edges has at least two Hamiltonian cycles.

More generally, $D_{n,k}$ shows that there are strong digraphs with n vertices and $(n(n-1)+2k-4)/2$ edges that contain no cycles of length k. We conjecture that this is the best possible for k even.

CONJECTURE 5 If k is even, and n is sufficiently large, then every strong digraph with n vertices and $(n(n-1)+2k-2)/2$ or more edges contains a k-cycle.

The paper [13] contains various generalizations of Theorem 8 and with the aid of these a result slightly weaker than Conjecture 5 is obtained.

THEOREM 10 (Häggkvist and Thomassen [13]) Let $k \geq 2$ be an even integer. If D is a strong digraph with n vertices and $(n(n-1)+(k-1)(k-2)+2)/2$ or more edges, then D contains a k-cycle.

There are digraphs of order n and with $n^2/2$ edges that contain no odd cycles (namely complete bipartite digraphs). This shows that Conjecture 5 and Theorem 10 are not valid for k odd. However, for k odd, we have

THEOREM 11 (Häggkvist and Thomassen [13]) A strong digraph

with $n \geq (k-1)^2$ vertices and more than $n^2/2$ edges contains a cycle of length k .

For k odd, Theorem 11 is best possible except that probably the condition $n \geq (k-1)^2$ can be weakened.

In order to determine the number of edges needed in a digraph of order n (which is not necessarily strong as it was assumed in the preceding discussion) to ensure a cycle of a given length k , we consider the union of s complete digraphs $H_1, H_2, \ldots, H_s$ each of order $\leq k-1$ such that the total number of vertices is n and such that each H_i , except possibly one, has order $k-1$. We then add all edges xy , where $x \in V(H_i)$, $y \in V(H_j)$, $i < j$. Let $g(n,k)$ denote the number of edges in the resulting digraph. It was proved in [13] that every digraph with n vertices and $(n(n-1)+(k-2)n)/2$ edges has a cycle of length k unless $n \equiv 0 \pmod{k-1}$ and the digraph has the structure described above. As pointed out in [27] the proof can be modified so as to give the following

THEOREM 12 (Häggkvist and Thomassen [13]) A digraph with n vertices and $g(n,k)$ or more edges contains a k-cycle unless it has the structure described above.

The weaker result that the assumption of Theorem 12 implies a cycle of length at least k was proved by Lewin [14].

The minimum number of edges needed in an undirected graph of order n to ensure a cycle of length k is known for k odd [2, 29] (for $n \geq 2k-2$ the number is $[n^2/4] + 1$; for $n \geq (k-1)^2$ this also follows from Theorem 11), but it is not known for k even. As we have seen the analogous problems for digraphs are solved almost completely. The minimum number of edges in an undirected graph of order n to ensure a cycle of length $\geq k$ was found by Erdös and Gallai [10] (for $n \equiv 1 \pmod{k-2}$)). For k fixed this is a linear function of n. The analogous result for digraphs is the above-mentioned result of Lewin [14]. However, it is not known how many edges a strong digraph with n vertices and no cycle of length $\geq k$ $(k \geq 4)$ can have. Put $s = [(k-1)/3]$ and consider the complete bipartite graph $K_{[(n-s)/2],[(n-s+1)/2]}$ and make it into an oriented graph with no path of length 2. Then add s vertices and let each of these dominate and be dominated by all other vertices. The resulting digraph is strong (it is in fact s-connected), it follows easily that it has no cycle of length $> 3s$, and it has $h(n,k) = [(n-s)^2/4] + s(2n-s-1)$ edges.

CONJECTURE 6 For each $k \geq 4$ there is a constant n_k such that every strong digraph with $n \geq n_k$ vertices and $h(n,k)$ or more edges has a cycle of length $\geq k$ unless the digraph has the structure described above.

The above-mentioned result of Erdös and Gallai [10] on cycles of length $\geq k$ in undirected graphs is based on Dirac's result that every 2-connected undirected graph with minimum degree $\geq k$ contains a cycle of length $\geq 2k$ or a Hamiltonian cycle. Bermond [1] and the author [27] independently conjectured that every 2-connected digraph with minimum indegree and outdegree $\geq k$ contains a cycle of length $\geq 2k$ or a Hamiltonian cycle. It is obvious that a digraph with minimum outdegree $\geq k$ contains a cycle of length $\geq k+1$ and with a little more effort one can prove the following

THEOREM 13 A 2-connected digraph with at least $k+2$ vertices and with minimum outdegree $\geq k$ contains a cycle of length $\geq k+2$.

As proved in [26] the above-mentioned conjecture of Bermond and the author is false and Theorem 13 is the best possible in the sense that there are infinitely many 2-connected digraphs with minimum indegree and outdegree $\geq k$ that contain no cycles of length $\geq k+3$. To see this we use the following simple construction which seems to be efficient in disproving other plausible conjectures on cycles in digraphs as well. Let H_1 and H_2 be disjoint 2-connected digraphs. Select vertices x_1, y_1 in H_1 and x_2, y_2 in H_2. Then form the union $H_1 \cup H_2$ and add the 4-cycle $x_1 \to x_2 \to y_1 \to y_2 \to x_1$. The resulting digraph H is 2-connected and if a cycle in this digraph contains at least two vertices of H_i $(i = 1 \text{ or } 2)$ then it contains at most one vertex of H_{3-i}. If H_1 and H_2 are both complete digraphs of order $k+1$, then H has no cycle of length $\geq k+3$. Also H is a counter-example to the conjecture of Bondy [4] that every k-connected digraph is Hamiltonian if it has the property that any set of $k+1$ vertices induces a cycle. If we delete from H the two edges joining x_1 and y_1 and the two edges joining x_2 and y_2, the resulting digraph is 2-connected and k-diregular. This shows that a result analogous to Corollary 1 with $n = 2k+2$ must include a list of infinitely many exceptional digraphs.

By iterating the construction above we can get an infinite family of k-diregular digraphs containing no cycle of length $\geq k+4$. However, we believe the following holds:

CONJECTURE 7 For each $k \geq 3$ there are only finitely many 2-connected k-diregular digraphs with no cycle of length $\geq k+3$.

As pointed out in [26] the construction above also enables us to construct, for each k, an infinite family of digraphs demonstrating that Theorem 13 is the best possible. If we exclude the existence of separating 4-cycles in the digraphs under consideration we may be able to improve a little on Theorem 13. However, examples in [26] show that for each k there are infinitely many 2-connected digraphs which have minimum indegree and outdegree $\geq k$, and which have no separating 4-cycles and which contain no cycles of length $\geq k+3$. These examples as well as the examples obtained by the construction above each contain two vertices that are not on a common cycle. We make the following conjecture:

CONJECTURE 8 If a digraph D has minimum indegree and outdegree $\geq k$ and any two vertices of D are on a common cycle, then D contains a cycle of length $\geq 2k$ or a Hamiltonian cycle.

It should be mentioned that the assumption of Conjecture 8 includes a condition that is not easy to check. A non-trivial necessary and sufficient condition for two vertices of a digraph to be on a common cycle is not known.

We conclude this paper with some remarks on Hamiltonian-connected digraphs. An undirected graph is *Hamiltonian-connected* if for any two distinct vertices of the graph there is a Hamiltonian path connecting them. In particular, every edge of the graph is contained in a Hamiltonian cycle. A digraph D is *strongly* (resp. *weakly*) *Hamiltonian-connected* if for any two vertices x and y of D, there is a Hamiltonian path from x to y and (resp. or) from y to x. Ore [20] proved that an undirected graph G of order n is Hamiltonian-connected if the sum of degrees of any two non-adjacent vertices is $\geq n+1$. In particular, G is Hamiltonian-connected if G has minimum degree $\geq (n+1)/2$. This was generalized to digraphs by Overbeck-Larish.

THEOREM 14 (Overbeck-Larish [21]) A 2-connected digraph with

n vertices and minimum degree $\geq n+1$ is weakly Hamiltonian-connected.

Overbeck-Larish (private communication) also suggested to generalize the above-mentioned result of Ore to digraphs. Since tournaments satisfy conditions of Ore-type, it is natural to investigate these digraphs first. In [25] it is described when a tournament has a Hamiltonian path between two prescribed vertices and from this result we get

THEOREM 15 ([25]) A tournament T with $n \geq 3$ vertices is weakly Hamiltonian-connected if and only if T is strong, for each vertex x of T, $T-x$ has at most two components, and T is not isomorphic to any of the tournaments of Figure 3.

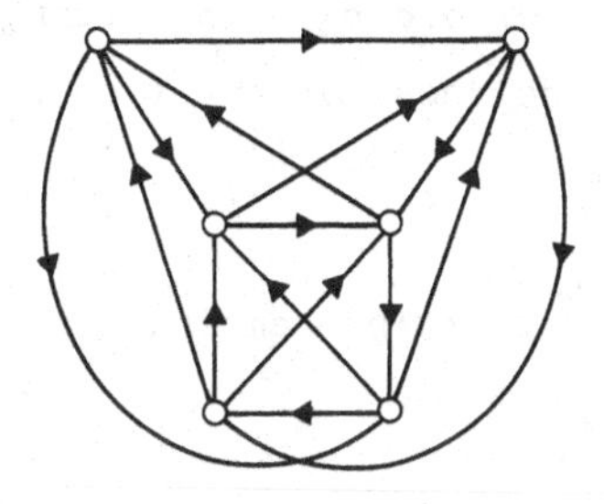
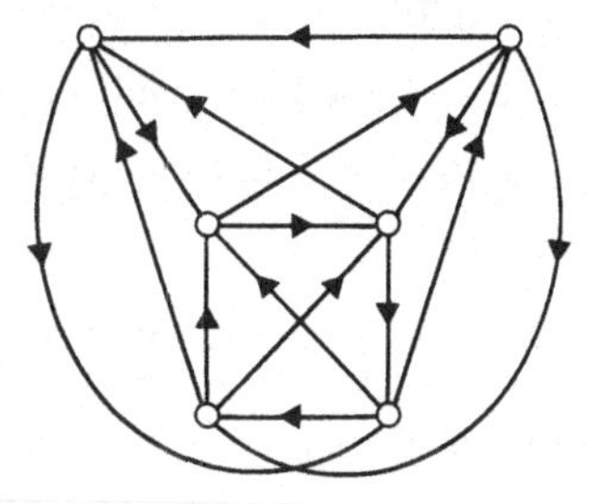

Figure 3: Two non-weakly-Hamiltonian-connected 2-connected tournaments

Strongly Hamiltonian-connected tournaments have not been completely characterized. However we have

THEOREM 16 ([25]) Any 4-connected tournament is strongly Hamiltonian-connected and every edge of a 3-connected tournament is contained in a Hamiltonian cycle. Moreover, there are infinitely many 3-connected tournaments that are not

Hamiltonian-connected and infinitely many tournaments containing an edge that is not contained in a Hamiltonian cycle.

If we want a result on strongly Hamiltonian digraphs analogous to Theorem 14, it is not sufficient to sharpen the degree condition a little. In [25] we describe for each $k \geq 1$ a 2-connected digraph D of order n and with minimum degree $n+k$ which is not strongly Hamiltonian-connected.

CONJECTURE 9 Every 3-connected digraph of order n and with minimum degree $\geq n+1$ is strongly Hamiltonian-connected.

Finally, the last part of Theorem 15 shows that if we want a result on strongly Hamiltonian-connected digraphs of Ore-type it is not enough to assume 3-connectedness.

CONJECTURE 10 A 4-connected digraph of order n is strongly Hamiltonian-connected if the sum of degrees for any two non-adjacent vertices is at least $2n+1$.

If true, Conjecture 10 extends part of Theorem 16.

REFERENCES

[1] Bermond, J.-C. (1975). *Cycles dans les graphes et G-configurations.* Thèse, Université Paris XI.

[2] Bondy, J.A. (1971). Large cycles in graphs. *Discrete Math.* 1, 121-132.

[3] Bondy, J.A. (1971). Pancyclic graphs I. *J. Combinatorial Theory* B 11, 80-84.

[4] Bondy, J.A. (1978). *Hamiltonian cycles in graphs and digraphs.* Preprint, University of Waterloo.

[5] Bondy, J.A. and Chvátal, V. (1970). A method in graph theory. *Discrete Math.* 15, 111-135.

[6] Bondy, J.A. and Thomassen, C. (1977). A short proof of Meyniel's theorem. *Discrete Math.* 19, 195-197.

[7] Camion, P. (1959). Chemins et circuits hamiltoniens des graphes complets. *C.R.Acad.Sci. Paris* 249, 2151-2152.

[8] Chvátal, V. (1972). On Hamilton's ideals. *J. Combinatorial Theory* B 12, 163-168.

[9] Dirac, G.A. (1952). Some theorems on abstract graphs. *Proc. London Math.Soc.* 2, 69-81.

[10] Erdös, P. and Gallai, T. (1959). On maximal paths and circuits of graphs. *Acta Math.Acad.Sci.Hung.* 10, 337-356.

[11] Ghouila-Houri, A. (1960). Une condition suffisante d'existence d'un circuit Hamiltonien. *C.R.Acad.Sci. Paris* 25, 495-497.

[12] Hakimi, S.L. and Schmeichel, E.F. (1974). Pancyclic graphs and a conjecture of Bondy and Chvátal. *J. Combinatorial Theory* B 17, 22-34.

[13] Häggkvist, R. and Thomassen, C. (1976). On pancyclic digraphs. *J. Combinatorial Theory* B 20, 20-40.

[14] Lewin, M. (1975). On maximal circuits in directed graphs. *J. Combinatorial Theory* B 18, 175-179.

[15] Meyniel, M. (1973). Une condition suffisante d'existence d'un circuit hamiltonien dans un graphe orienté. *J. Combinatorial Theory* B 14, 137-147.

[16] Moon, J.W. (1968). *Topics on tournaments.* Holt, Rinehart and Winston, New York.

[17] Müller, V. and Pelant, J. (1978). The number of Hamiltonian circuits. *J. Combinatorial Theory* B 24, 223-227.

[18] Nash-Williams, C.St.J.A. (1969). Hamilton circuits in graphs and digraphs. In *The Many Facets of Graph Theory.* Lecture Notes in Mathematics. Springer Verlag 110, 237-243.

[19] Ore, O. (1960). Note on Hamilton circuits. *Amer.Math. Monthly* 67, 55.

[20] Ore, O. (1963). Hamilton connected graphs. *J.Math. Pures Appl.* 42, 21-27.

[21] Overbeck-Larisch, M. (1976). Hamiltonian paths in oriented graphs. *J. Combinatorial Theory* B 21, 76-80.

[22] Overbeck-Larisch, M. (1977). A theorem on pancyclic oriented graphs. *J. Combinatorial Theory* B 23, 168-173.

[23] Pósa, L. (1962). A theorem concerning Hamiltonian lines. *Magyar Tud.Akad.Math. Kutató Int.Közl.* 7, 225-226.

[24] Thomassen, C. (1977). An Ore-type condition implying a digraph to be pancyclic. *Discrete Math.* 19, 85-92.

[25] Thomassen, C. Hamiltonian-connected tournaments. *J. Combinatorial Theory* B (to appear).

[26] Thomassen, C. Long cycles in digraphs. (to appear).

[27] Thomassen, C. (1976). *Paths and cycles in graphs.* Ph.D. Thesis, University of Waterloo.

[28] Tillson, T. A Hamiltonian decomposition of K^*_{2m}, $2m \geq 8$. *J. Combinatorial Theory* B (to appear).

[29] Woodall, D.R. (1972). Sufficient conditions for circuits in graphs. *Proc. London Math.Soc.* 24, 739-755.

9 · Colouring problems and matroids

Dominic Welsh

1. INTRODUCTION

One of the attractions of matroid theory is that it usually provides several different natural settings in which to view a problem. The theme of this paper is the use of matroids to relate problems about colourings and flows in graphs with problems of projective geometry.

O. Veblen in 1912 was the first to attempt to settle the four colour conjecture by geometrical methods. In the same year G.D. Birkhoff tried to settle it by an enumerative approach. Despite, or perhaps because of, their lack of success their work led to some startling extensions, notably by W.T. Tutte. Following on from Veblen's work, Tutte in 1966 formulated a fascinating geometrical conjecture - namely that there were just three minimal geometrical configurations which had a non empty intersection with each coline of PG(n,2) (he called them tangential 2-blocks). As a result of his work on enumerating polynomials Tutte in 1954 was led to make a series of equally fascinating conjectures about flows in directed networks. One of these was recently proved by F. Jaeger in 1975, when he gave a very elegant proof that every bridgeless graph has an 8-flow - equivalently every bridgeless graph is the union of three Eulerian subgraphs. Even more recently a remarkable advance towards the solution of Tutte's tangential block conjecture was made by P.D. Seymour when he showed that the only new tangential blocks must be cocycle matroids of graphs. This reduces a seemingly intractable geometrical question to the conceptually easier problem of classifying those graphs which cannot be covered by two Eulerian subgraphs. This area of combinatorics fits very easily into matroid theory since the latter provides a natural framework from which it is convenient to use either geometric or graph theoretic arguments. The key concept is the idea of extending the chromatic polynomial of a graph to arbitrary matroids. For simple matroids this becomes the characteristic polynomial of the corresponding geometric lattice and several important problems in combinatorics are known to reduce to finding the zeros of this polynomial. In this paper we survey the recent work done in this area.

First our notation; a graph G with vertex set $V(G)$ and edge set $E(G)$ will be allowed to have loops and multiple edges. A *cycle* is a simple closed path and a *cocycle* is a minimal set of edges whose removal increases the number of connected components of G. A *bridge* is a cocycle consisting of a single edge. G is *k-edge connected* if each cocycle has cardinality at least k. For any edge e, G'_e and G''_e denote respectively the graphs obtained from G by deleting and contracting the edge e. A *cotree* of G is the complement in $E(G)$ of the edge set of any spanning forest of G. G is *Eulerian* if its edge set is the union of disjoint cycles. The other graph terminology is standard, see for example the books of Berge (1973) or Bondy and Murty (1976). Throughout M will denote a matroid on S with rank function ρ. If $e \in S$ we write M'_e, M''_e respectively for the matroid obtained from M by deleting or contracting the element e from M. A *minor* is any matroid N on a subset T of S obtained from M by a sequence of deletions and contractions. We write $M|T$ and $M.T$ respectively to denote the *restriction* and *contraction* of M to T. A matroid is *graphic* if it is the *cycle matroid* of some graph G, that is if its circuits are the edge sets of cycles of G. It is *cographic* if it is the *cocycle matroid* of some graph G, that is if its circuits are the edge sets of cocycles of G. We denote these matroids by $M(G)$ and

$M^*(G)$ respectively and it is well known that for a given graph they are dual matroids. In general the *dual* of a matroid M will be denoted by M^*.

A *binary* matroid is one which is representable over $GF(2)$ in such a way as to preserve independence, in other words M is binary if it is embeddable in the projective space $PG(n,2)$. M is *regular* if it is representable over every field. We use $V(r,q)$ to denote the vector space of rank r over $GF(q)$ and use *k-flat* to denote both a subspace of rank k in $V(r,q)$ or the corresponding set of points in projective space. The *Fano matroid* ($\equiv PG(2,2)$) is denoted by F_7, the complete graph on n vertices is K_n, and the Petersen graph is denoted by P_{10}. Any remaining matroid terms not defined as encountered will be found in Welsh (1976).

2. TUTTE-GROTHENDIECK INVARIANTS

Brylawski (1972) introduced the idea of a Tutte-Grothendieck (T-G)-invariant for matroids as follows. An *invariant* is a function f on the set of matroids and taking values in a commutative ring such that M isomorphic to $N \Rightarrow f(M) = f(N)$ It is a *(T-G)-invariant* if in addition it satisfies, for any e which is not a loop or coloop:

$$f(M) = f(M'_e) + f(M''_e) \qquad (1)$$

$$f(M_1 + M_2) = f(M_1)f(M_2) \quad . \qquad (2)$$

It is easy to check that examples of T-G-invariants are:

the number $b(M)$ of bases of M ; (3)

the number $i(M)$ of independent sets of M ; (4)

the number $sp(M)$ of spanning sets of M . (5)

Brylawski's main result is the following theorem.

THEOREM 1 *Any Tutte-Grothendieck invariant is uniquely determined by its values on the two single element matroids.*

More precisely there exists a polynomial $T(M;x,y)$ in two variables such that if f is a T-G invariant and if $f(M_0) = x$, $f(M_0^*) = y$ where M_0 is the single element matroid of rank 1, then

$$f(M) = T(M;x,y) \quad .$$

The polynomial $T(M;x,y)$ is known as the *Tutte polynomial* of the matroid M. It is most easily defined by its relation to the *(Whitney) rank generating function* $R(M;x,y)$ by

$$R(M;x,y) = T(M;x-1,y-1)$$

where

$$R(M;x,y) = \sum_{A \subseteq S} x^{\rho S-\rho A} y^{|A|-\rho A} \quad .$$

The *chromatic polynomial* $P(M;\lambda)$ of the matroid M is defined by

$$P(M;\lambda) = (-1)^{\rho S} T(M;1-\lambda,0) \quad . \tag{6}$$

When M is the cycle matroid of a graph G, we know from elementary graph theory that the chromatic polynomial $P(G;\lambda)$ of the graph G satisfies

$$P(G;\lambda) = P(G_e';\lambda) - P(G_e'';\lambda)$$

when e is neither a bridge nor loop of G. The analogue

of (2) is trivial and it is now straightforward to see that

$$\lambda^{k(G)} P(M(G);\lambda) = P(G;\lambda)$$

where $k(G)$ denotes the number of connected components of G. Accordingly if we define the *chromatic number* $\chi(M)$ of a matroid by

$$\chi(M) = \inf_{k \in Z^+} k : P(M;k) > 0 \quad ,$$

then when M is graphic, the chromatic number of M agrees with the chromatic number of the underlying graph. Incidentally this shows that any two graphs whose cycle matroids are isomorphic have the same chromatic number. For a derivation and proof of all these results see Welsh (1976) chapter 15.

The bulk of this paper is concerned with the chromatic polynomial of M, but since it is so closely related to the Tutte polynomial we first survey some recent results in this area. First note that a *generalised TG invariant* cen be defined as an invariant f of matroids satisfying (2) and the generalised version of (1), namely:

for any fixed non-zero real numbers a and b,
and any matroid M,

$$f(M) = af(M'_e) + bf(M''_e) \quad . \tag{7}$$

Oxley and Welsh (1978) show that (2) and (6) together with the values x,y which f takes on the single element matroids consisting of a single coloop and a single loop, uniquely determine f by the relation

$$f(M) = a^{|S|-\rho S}\, b^{\rho S}\, T(M; b^{-1}x, a^{-1}y) \quad . \tag{8}$$

This is applied to problems involving random graphs and percolation theory; it is also shown that there is no way of extending the ideas underlying this theory to structures which do not arise in some way from matroids.

Greene (1976) has shown the intimate relationship between the weight polynomial of a linear code U over $GF(q)$ and the Tutte polynomial of the matroid of the column vectors of any matrix generating U. He has used the identity

$$T(M;x,y) = T(M^*;y,x) \tag{9}$$

to rederive the MacWilliams' identities for linear codes.

For planar graphs Martin (1978) obtained a very curious interpretation of their Tutte polynomial evaluated at $(-1,-1)$. This has been extended to binary matroids by Rosenstiehl and Read (1978); they define a *bicycle* to be a set of elements which is both a disjoint union of circuits and a disjoint union of cocircuits. They show that if M is binary on n elements then

$$T(M;-1,-1) = (-1)^{|S|}(-2)^q$$

where q is the dimension of the bicycle space.

Zaslavsky (1975) used Tutte-Grothendieck arguments to obtain formulae which count the regions of E^d partitioned by a given arrangement of hyperplanes. For an account of this and related topics see also Brylawski (1976) where similar matroid arguments are used to count the number of complementary subsets of points of a set S whose convex hulls intersect. In fact the number of regions in any arrangement of hyperplanes can be obtained as an evaluation of a suitable matroid.

Another evaluation of the Tutte polynomial was found by Stanley (1973) who showed that for a graph G, $P(G;-1)$

counted the number of acyclic orientations. Brylawski and Lucas (1976) extended this to regular matroids and Las Vergnas (1979) has recently shown that this result can be extended to the much wider class of oriented matroids introduced by Bland and Las Vergnas. Las Vergnas proves:

If M is an oriented matroid on S and $_{\bar{A}}M$ denotes the oriented matroid obtained from M by reversing the signs on A then $P(M;-1)$ counts the number of subsets A of S such that $_{\bar{A}}M$ is acyclic. (10)

Similarly, if we call a set *totally cyclic* if each element is a member of a directed cycle, Las Vergnas dualises (10) to:

In an oriented matroid M on S, $P(M^*;-1)$ counts the number of subsets A of S such that $_{\bar{A}}M$ is totally cyclic. (11)

In another paper Las Vergnas has extended the ideas underlying (T-G)-invariance and hence defined a polynomial $T(M,N)$ associated with two matroids M and N which are related by a strong map. He shows that basic properties of the usual Tutte polynomial generalise to $T(M,N)$.

3. THE FLOW POLYNOMIAL

Suppose G is a graph, and that ω is any orientation of the edges of G. If H is any Abelian group we say that an injection $\phi: E(G) \to H\backslash\{0\}$ is an *H-flow* if for each vertex $v \in G$

$$\sum_{\partial^+(v)} \phi(e) - \sum_{\partial^-(v)} \phi(e) = 0 \tag{1}$$

where $\partial^+(v), \partial^-(v)$ denote respectively the edges of G which are oriented into (out of) v in the orientation ω. Note that what we call an H-flow is called by most authors a *nowhere zero H-flow*. If ω and ω' are any two orientations of G then there is an obvious one-to-one correspondence between flows in ω and ω'. Thus the number $N(H;G)$ of H-flows in G is independent of its orientation and is a function only of H and G. Secondly we note that if G is not 2-connected, and has cyclic components say G_1, G_2,

$$N(H;G) = N(H;G_1)N(H;G_2) \quad . \tag{2}$$

Slightly less obvious is that if e is not a loop or bridge of G,

$$N(H;G) = -N(H;G'_e) + N(H;G''_e) \quad . \tag{3}$$

<u>Proof</u> Consider any orientation ω of G and let F be the set of functions $\phi: E(G) \to H$ which satisfy (1). Partition F into $F_1 \cup F_2$ where F_1 are those functions ϕ which are H-flows, that is which are non-zero on e, and $F_2 = F \setminus F_1$. Then any member of F_1 is clearly an H-flow on G''_e with the induced orientation, and all such flows can be seen to arise in this way. Similarly all members of F_2 are seen to correspond exactly to the H-flows on G'_e. This proves (3).

Hence we have shown that on the class of cographic matroids, $N(H;G)$ is a generalised TG invariant. But clearly if M_0 is a coloop $N(H;M_0) = 0$ while if M_1 is a loop, $N(H;M_1) = O(H) - 1$ where $O(H)$ is the order of H. Hence an immediate application of (2.8) gives the following result

$$N(H;G) = (-1)^{|S|-\rho(S)} T(M(G);0,O(H)-1) \quad . \tag{4}$$

Thus by (2.6) we have the interesting result first noted by Tutte in 1954,

THEOREM 1 *The number of H-flows on a graph G depends only on the order of the group H and is given by*

$$N(H,G) = P(M^*(G);O(H)) \quad . \tag{5}$$

Because of (5), we can, without ambiguity, use the terminology G has a *k-flow* to mean that G has an H-flow for all (or any) group H of order k . In 1954 Tutte first made his conjecture that there existed an integer k such that every bridgeless graph has a k-flow. This conjecture was settled in 1975 when Jaeger gave a very elegant proof of the following theorem

THEOREM 2 *Every bridgeless graph has an 8-flow.*

Sketch proof Let K be any cotree of G , that is $E(G)\backslash K$ is a spanning tree. For each $e \in K$ let C_e denote the fundamental circuit of e in the tree $T = E\backslash K$. Let $\phi_e: E(G) \to Z_2$ be defined by $\phi_e(f) = 1$ if and only if $f \in C_e$. Then clearly whatever orientation we have on G , ϕ_e is a Z_2-flow. Let ϕ_K be defined by

$$\phi_K = \sum_{e \in K} \phi_e \quad .$$

Then ϕ_K is also a Z_2-flow. Hence if we can find three cotrees K_1, K_2, K_3 such that $K_1 \cup K_2 \cup K_3 = E(G)$ then the related $(Z_2 \times Z_2 \times Z_2)$-flow is nowhere zero on $E(G)$. But by using basic graph theory together with the well known matroid result of Edmonds that for any matroid M on S with rank function ρ , S can be covered by t bases of M if and only if $t\rho(A) \geq |A|$ for all $A \subseteq S$, Jaeger shows that any bridgeless three edge connected graph can be covered by three cotrees. Since $Z_2 \times Z_2 \times Z_2$ has order 8 the theorem follows for graphs with no cocycle of size 3. The remaining case when G has a 2-cocycle is fairly straightforward; it involves breaking up the graph G at the cocycle, finding flows on each of the two parts and then sticking the two parts (with their flows) back together to get a flow on G.

Immediate consequences of this result and its method of proof are the following statements.

COROLLARY *Every bridgeless graph G is the union of 3 Eulerian subgraphs.*

Proof Consider a $(Z_2 \times Z_2 \times Z_2)$-flow ϕ on G. For any edge e, write $\phi(e) = (\phi_1(e), \phi_2(e), \phi_3(e))$. Then decompose $E(G)$ into three Eulerian graphs E_1, E_2, E_3 where an edge e is a member of E_k if and only if $\phi_k(e) = 1$ for $k = 1,2,3$.

A second conjecture of Tutte suggests that Jaeger's theorem is not the best possible:

Tutte's 5-flow conjecture Every bridgeless graph has a 5-flow.

If true, this could not be improved since Tutte proved in 1954:

The Petersen graph P_{10} has no 4-flow. (6)

It is not difficult to show that Tutte's 5-flow conjecture is true if it can be shown that all bridgeless cubic graphs have a 5-flow, and for cubic graphs Minty (1967) shows that a cubic bridgeless graph has a 4-flow if and only if its edges may be coloured with 3 colours so that no pair of incident edges are the same colour. Hence the following conjecture by Tutte (1969):

If a cubic bridgeless graph G has chromatic index 4 then G contains a subgraph contractible to the Petersen graph P_{10} : (7)

is almost equivalent to the *4-flow conjecture*

If a bridgeless graph has no 4-flow then it has a subgraph contractible to P_{10} . (8)

Using the same argument as he used to prove his main theorem Jaeger also shows that if G is 4-edge connected then it is the union of 2 cotrees and hence a simple modification to his proof gives:

THEOREM 3 *If a graph G is 4-edge connected it has a 4-flow.*

Also we know by the 4-colour theorem of Appel and Haken that $P(M:4) > 0$ whenever M is the cycle matroid of a planar graph. Hence by duality we have:

Any bridgeless planar graph has a 4-flow. (9)

Thus we can restrict attention to cubic, bridgeless, non-planar graphs in any attempt to settle Tutte's 5-flow conjecture.

4. THE COLOURING PROBLEM AND PROJECTIVE GEOMETRY

Consider an injection $\psi : E(G) \to PG(n,2)$ which preserves independence, and which is such that ψX is linearly independent in $PG(n,2)$ if and only if X is independent in $M(G)$. Suppose also that $PG(n,2)$ and $M(G)$ have the same rank $n+1$. If C^* is any cocycle of G then $E\backslash C^*$ is a hyperplane of $M(G)$ and hence $E\backslash C^*$ is contained in a unique hyperplane H of $PG(n,2)$. Consider now any 4-colouring of the vertices of G. We may take the colours to be the four vectors of $PG(2,2)$. Let U_1 be the set of vertices whose colour vector has a first component equal to one and U_2 those vertices for which the second component is 1. Any edge must join edges of different colours and hence must belong to ∂U_1 or ∂U_2. But since ∂U_1 is the set of edges joining U_1 to $V\backslash U_1$, it is the union of cocycles. Thus if G has a 4-colouring we can write $E = D_1^* \cup D_2^*$ where D_i^* is the union of disjoint cocyles (we shall call it a *coboundary*). Conversely it can be shown that if there exist a pair of coboundaries D_1^*, D_2^* whose union is $E(G)$ then the graph G is 4-colourable. For an account of this see Tutte (1966).

Hence the planar graph G has a 4-colouring iff there exist hyperplanes H_1, H_2 of the embedding space V such that $(E\backslash H_1) \cup (E\backslash H_2) = E$. In other words G is 4-colourable if and only if there exists an embedding $\psi : M(G) \to PG(n,2)$ $(n = \rho(M(G)))$ and hyperplanes H_1, H_2 of $PG(n,2)$ such that H_1, H_2 and $\psi(E)$ are disjoint.

More generally it can be shown, see Welsh (1976, chapter 15) that a graph has a 2^k-colouring if and only if there exists an embedding $\psi : E(G) \to PG(n,2)$ and hyperplanes $H_1, \ldots, H_k$ of $PG(n,2)$ such that $\psi(E) \cap H_1 \cap \ldots \cap H_k = \phi$.

Thus we have a natural extension of the colouring problem to any matroid which is representable over a field $GF(q)$. In particular:

A binary matroid M of rank r on S has chromatic number 4 if and only if it can be embedded in $V = V(r,2)$ in such a way that there exist hyperplanes H_1, H_2 of V such that

$$H_1 \cap H_2 \cap S = \phi \quad . \tag{1}$$

But a pair of hyperplanes of $V(r,2)$ intersect in an $(r-2)$-flat and hence the more general geometrical version of the 4-colour theorem is:

Problem Find the minimal sets of points in $PG(n,2)$ which meet every $(n-2)$-flat.

This was a special case of the following problem studied by Tutte in 1966. A *k-block* is a set F of points in $V(r,2)$ whose rank is $> k$ and which intersects every $(r-k)$-flat of $V(r,2)$. A k-block F is *minimal* if no proper subset of F is also a k-block. When $k = 1$, the situation is straightforward as Tutte proves with an easy geometrical argument:

A minimal 1-block is any dependent set of points $x_1, \ldots, x_{2n+1}$ of $V(r,2)$ such that any proper subset is linearly independent. (2)

In other words:

The minimal 1-blocks of $V(r,2)$ are the odd circuits. (3)

However when $k \geq 2$ the situation is more complicated. For example it will be seen that any edge critical 5-chromatic graph is a minimal 2-block. Similarly F_7 is a non-graphic minimal 2-block. Indeed if $S(M,N)$ denotes the series connection of two matroids (a *series connection* is a matroid operation defined by Brylawski (1973) and corresponds to the Hajos union of two graphs), then Oxley (1978) proves:

If M,N are two minimal k-blocks then their series connection $S(M,N)$ is also a minimal k-block. (4)

Regarded as a matroid it is easy to find examples of minimal blocks which contain minimal blocks as minors (that is as contraction minors). This lack of 'minor minimality' underlies Tutte's definition of a tangential block.

Let C be a non-null subset of a k-block B. We define a *tangent* of C in B to be any (r-k) subspace in $V(r,2)$ which contains all the points of C but no point of B which is independent of them. We call B a *tangential k-block* if every non-null subset of B, of rank not exceeding $r-2$ has a tangent in B.

First note:

Every tangential k-block is minimal. (5)

Proof Let B be a tangential k-block which is not minimal. Then there exists $x \in B$ such that $B\backslash x$ is a k-block. But each tangent of $\{x\}$ is an (r-k)-space which does not meet $B\backslash x$, which is a contradiction.

A second result of Tutte (1966) is the following.

Let B be a k-block in $V(r,2)$. Then B is not tangential if and only if it has a closed subset C such that the projection of B from C transforms B into another k-block. (6)

Now let M be a matroid of rank r on S which is embedded in $V(r,q)$. The *critical exponent* or (number) $c(M;q)$ is defined to be the minimum number c such that there exists a c-tuple $(H_1,\ldots,H_c)$ of hyperplanes of $V(r,q)$ such that $H_1 \cap \ldots \cap H_c \cap S = \phi$. Its value is given by the first integer c such that $P(M;q^c) > 0$ or by ∞ if M has a loop.

In fact for any positive integer k, $P(M;q^k)/(q-1)^k k!$ is the number of k-sets of hyperplanes whose intersection with S is empty. This is immediate from the definition of critical exponent in terms of distinguishing hyperplanes by Crapo and Rota (1970). From this is it also evident that $c(M;q) = 1$ if and only if M can be embedded in the affine geometry $AG(n,q)$. Since $H_1 \cap \ldots \cap H_c$ is an (r-c)-flat of $PG(r-1,q)$ it is clear that we can reformulate Tutte's theory of k-blocks as follows.

A *k-block* is a simple binary matroid M with critical exponent $c(M;2) > k$. It is a *minimal k-block* if in addition $c(M_e';2) = k$ for each element $e \in S$. Then using the equivalence between geometric projection and matroid contraction we see that a matroid M is a *tangential k-block* if M is a k-block but no proper loopless minor of M is a k-block.

Thus we have:

There is only one tangential 1-block, namely the triangle $M(K_3)$. (7)

For each $k \geq 2$, $PG(k,2)$ is a tangential k-block. (8)

For each positive integer k, the cycle matroid of the complete graph K_m, $m = 2^k + 1$ is a tangential k-block. (9)

Now the reader will recall from §3 that if P_{10} is the Petersen graph then $\chi(M^*(P_{10})) = 5$ because P_{10} has no 4-flow and is minimal with this property. Thus $M^*(P_{10})$ is another tangential 2-block. A third tangential 2-block is the Fano matroid F_7 and the main theorem of Tutte (1966) is:

THEOREM 1 *The only tangential 2-blocks of rank* ≤ 6 *are* F_7, $M(K_5)$ *and* $M^*(P_{10})$.

Moreover in 1966 Tutte made the following, still open, conjecture:

<u>Conjecture</u> The only tangential 2-blocks are F_7, $M(K_5)$ and $M^*(P_{10})$. (10)

Until recently little progress has been made on this conjecture though Datta (1976), by some intricate geometrical arguments, has extended Tutte's theorem by proving

THEOREM 2 *There are no tangential 2-blocks of rank* 7 .

The next section is devoted to an account of a beautiful theory developed recently by P.D. Seymour, which as we shall see reduces the apparently very difficult geometrical problem of classifying all tangential 2-blocks to the purely graph theoretic problem of deciding whether the only contraction minimal graph without a 5-flow is P_{10} .

5. SEYMOUR'S THEORY OF SPLITTERS

Let $\mathcal{F}$ be a class of binary matroids which is closed under the taking of minors. Denote by $\mathcal{F}^*$ the class of duals of the matroids in $\mathcal{F}$. We say that $N \in \mathcal{F}$ is *compressed* in F if:

(a) N is simple, that is N has no circuit of size ≤ 2 .

(b) If $M \in F$ and is such that $M'_e = N$, then e is a loop, a coloop or is parallel to some other element of M .

The matroid $N \in F$ is a *splitter* for F if every $M \in F$ with a minor isomorphic to N has a 1- or 2-separation unless $M = N$.

A matroid M on S has a *k-separation* if there exists $A \subseteq S$, $|A| \geq k$, $|S \backslash A| \geq k$, and

$$\rho A + \rho(S \backslash A) \leq \rho S + (k-1) .$$

Now 1-separation corresponds exactly to a graph not being cyclically connected, while M(G) has a 2-separation but not a 1-separation if and only if G is 2-connected but has a pair of disconnecting vertices. In either case knowledge of a splitter for a class gives considerable leverage to inductive proofs since we either know the matroid exactly or can break it up into smaller submatroids.

Seymour's first key theorem is the following:

THEOREM 1 *Suppose that* F *is a class of binary matroids which is closed under minors and isomorphism and that* $N \in F$ *.* *If* N *is non-null, connected and satisfies the conditions:*

(i) N *is compressed in* F *;*

(ii) N^* *is compressed in* F^* *;*

(iii) N *is not isomorphic to the cycle matroid* $M(W_n)$ *of the wheel graph* W_n *for any* $n \geq 3$ *;*

then N *is a splitter for* F *.*

From this, the reader can, with sufficient time, check

that the following are examples of splitters.

F_7^* is a splitter for the class of binary matroids with no F_7 minor. (1)

If $\mathcal{F}$ is the class of regular matroids then the matroid R_{10}, consisting of the 10, 5-vectors over GF(2) with exactly 2 non-zero coordinates, is a splitter for $\mathcal{F}$ (see Bixby (1977)). (2)

To fully appreciate Seymour's results we need one more set of definitions.

If M_1, M_2 are binary matroids on S_1 and S_2 respectively we define $M_1 \Delta M_2$ to be the matroid on the symmetric difference $S_1 \Delta S_2$ and with cycles all subsets of $S_1 \Delta S_2$ of the form $C_1 \Delta C_2$ where C_i is a cycle of M_i. (A *cycle* of a matroid is a union of disjoint circuits.)

(a) When $S_1 \cap S_2 = \phi$ and $|S_1|, |S_2| < |S_1 \Delta S_2|$, we call $M_1 \Delta M_2$ a *1-sum*.

(b) When $S_1 \cap S_2 = \{z\}$ say and z is not a loop or coloop of M_1 or M_2 and $|S_1|, |S_2| < |S_1 \Delta S_2|$, we say $M_1 \Delta M_2$ is a *2-sum* of M_1 and M_2.

(c) When $|S_1 \cap S_2| = 3$ and $S_1 \cap S_2 = Z$ say where Z is a circuit of M_1 and M_2, and Z contains no cocircuit of either M_1 and M_2 and $|S_1|, |S_2| < |S_1 \Delta S_2|$ then we say $M_1 \Delta M_2$ is a *3-sum* of M_1 and M_2.

In each case we call M_1 and M_2 the *parts* of the sum.

Now a 1-sum is just the direct sum of Welsh (1976), the 2-sum and 3-sum are the matroid operations corresponding to 'sticking' two graphs together by an edge or triangle respectively and then deleting the edge or triangle in question. In either case if we have a matroid M which is the 2- or

3-sum of other matroids, then we have a convenient way of breaking it up into these smaller structures.

Seymour's main theorem is the following remarkable characterisation of regular matroids.

THEOREM 2 *If* M *is a regular matroid it is the* 1-, 2- *or* 3-*sum of graphic matroids, cographic matroids, and copies of the* 10 *element matroid* R_{10} .

The power of this result has not yet been fully exploited; however, as one major example we show how it gives 'as a corollary' a major result which shows that Tutte's 4-flow conjecture is equivalent to his tangential block conjecture.

THEOREM 3 *Any tangential* 2-*block other than* $M(K_5)$ *and* F_7 *must be the cocycle matroid of a graph.*

In other words there exists a tangential 2-block other than F_7 , $M(K_5)$ and $M^*(P_{10})$ if and only if there is a bridgeless graph G not containing a subgraph contractible to P_{10} which has no 4-flow.

<u>Sketch proof of Theorem 3</u> First consider a tangential 2-block M which is not $M(K_5)$, F_7 or $M^*(P_{10})$. If it were graphic there would exist a graph G which was not contractible to K_5 but which was not 4-colourable. This would contradict Hadwiger's conjecture for the case $n = 5$, which by Wagner's theorem showing the equivalence of Hadwiger's 5-chromatic conjecture with the 4-colour theorem of Appel and Haken, we know to be true. Hence the only tangential 2-block which is graphic is $M(K_5)$. Now consider the existence of a non-regular tangential 2-block M_0 . Since M_0 is not regular it must contain either F_7 of F_7^* as a minor. But because F_7 is a tangential 2-block,

by minimality, this minor must be F_7^*. Hence by (1), since F_7^* is a splitter for binary matroids with no F_7 minor, we know that either $M_0 = F_7^*$ or M_0 has a 2-separation. It is not difficult to show that a tangential 2-block cannot have a 2-separation. Hence by (3), $M_0 = F_7^*$. But $\chi(F_7^*) = 2$ and hence F_7^* is not a tangential 2-block. So we have shown that there are no non-regular tangential 2-blocks. It remains to show that there exists no tangential block which is regular but neither graphic nor cographic.

Suppose there did exist such a matroid M. Then by Theorem 2 it must be possible to express M as a 1-, 2- or 3-sum of graphic or cographic matroids or copies of R_{10}. A straightforward induction argument shows that this is impossible unless M is cographic and hence the only tangential blocks which are not cographic are F_7 and $M(K_5)$.

We close by remarking that the discovery of a splitter for a class of matroids usually gives great insight into the structure of the class. Two other splitters discovered by Seymour are the following:

$M(K_5)$ is a splitter for the class of regular matroids with no minor isomorphic to $M(K_{3,3})$. (3)

If W_8 denote the cubic graph with eight vertices $a_1, a_2, a_3, a_4, b_1, b_2, b_3, b_4$ and edge set $\{(a_i b_i): 1 \le i \le 4\} \cup \{(a_i a_{i+1}): 1 \le i \le 3\} \cup \{(b_i b_{i+1}): 1 \le i \le 3\} \cup (a_4 b_1) \cup (a_1 b_4)$ used by Wagner (1964) then:

$M(W_8)$ is a splitter for the class of regular matroids without $M(K_5)$ as a minor. (4)

6. FURTHER RESULTS ON CHROMATIC POLYNOMIALS

We conclude with a brief survey of known results on chromatic numbers and polynomials for general matroids.

First we emphasize that unlike the chromatic number of a graph the chromatic number of a matroid has some highly undesirable properties. Consider the chromatic polynomial of the projective geometry $PG(n,q)$:

$$P(PG(n,q);\lambda) = \prod_{i=0}^{n} (\lambda - q^i) \tag{1}$$

and so we have

$$\chi(PG(n,2)) = \begin{cases} 3 & n \text{ odd,} \\ 5 & n \text{ otherwise.} \end{cases}$$

Hence by considering $M(K_n)$ which is of rank $n-1$, binary and hence embeddable in $PG(n-2,2)$, we have an example of a matroid M and a restriction minor $M|T$ such that $\chi(M|T) - \chi(M)$ can be arbitrarily large.

By considering (1) we also notice that for even a binary matroid M it is not true that $P(M:n) \geq 0$ for each non-negative integer n. This is however an important property of graphic and cographic matroids since in each case $P(M:n)$ is a counting function, enumerating the number of n-colourings or n-flows as the case may be. Crapo (1969) generalizes this to regular matroids as follows.

Recall from Welsh (1976, chapter 10) that a regular matroid is orientable in the following sense. Consider the circuit and cocircuit incidence matrices of M on S, C and D defined by

$$C(X,e) = \begin{cases} 1 & \text{if } e \in X \text{ where } X \text{ is a circuit of } M, \\ 0 & \text{otherwise;} \end{cases}$$

$$D(x,e) = \begin{cases} 1 & \text{if } e \in X \text{ where } X \text{ is a cocircuit of } M, \\ 0 & \text{otherwise.} \end{cases}$$

Since M is regular there exists an assignment of signs ± 1 to the entries of C and D forming matrices C_0, D_0 such that each row of C_0 is orthogonal to each row of D_0 over R. Let the entries of the ith row D be $d_i(e)$, $e \in S$, $i = 1, \ldots, m$. If H is any Abelian group an *H-coboundary* is a function $f : S \to H$ of the form

$$f(e) = \sum_{i=1}^{m} d_i(e) h_i$$

where $h_i \in H$. An H-coboundary f is *nowhere zero* if $f(e) \neq 0$, $e \in S$. Then one of the main results of Crapo is the following:

THEOREM 1 *If* M *is a regular matroid on* S *and* H *is any Abelian group of order* n *, then the number of nowhere zero* H-*coboundaries on* M *is given by* $P(M;n)$.

Using this Lindström (1978) is able to prove:

If M is regular on S and S is the union of co-circuits each of cardinality less than k then

$$\chi(M) \leq k \quad . \tag{2}$$

This can be regarded as an analogue of Brooks' theorem bounding the chromatic number of a graph.

More generally Lindström is also able to show:

If M on S is representable over $GF(q)$ and S is the union of corcircuits each of cardinality less than q^k then

$$c(M;q) \leq k \quad . \tag{3}$$

For regular matroids we also have:

If M is regular and $P(M;n) > 0$ then
$P(M;n+1) > 0$. (4)

Proof It follows from the unimodular property of regular matroids that there is a nowhere zero Z_n-coboundary on M if and only if there is a nowhere zero Z-coboundary on M with all values in $[1-n,n-1]$.

As we have seen, the statement corresponding to (4) does not hold for binary matroids.

Various attempts have been made to obtain bounds for the chromatic number of a general non-regular matroid analogous to other bounds for the chromatic number of a graph. We list these below. The first due to Heron was:

If M is simple of rank r , then
$\chi(M) \leq \rho^*(M) + 2$. (5)

As several well known upper bounds for the chromatic number of a graph are stated in terms of the vertex degrees of the graph, a natural matroid generalisation is to obtain bounds on $\chi(M)$ which are obtained by substituting 'cocircuit size' for 'vertex degree' in the graph result. Oxley (1978) obtained another matroid version of Brooks' theorem by proving

If M is a simple matroid then
$$\chi(M) \leq 1 + \max_{C \in C^*(M)} |C|$$
where $C^*(M)$ is the set of cocircuits of M . (6)

A stronger result of Oxley can be obtained for regular matroids: let $R(M)$ be the set of simple restriction minors of M , then

If M is a simple regular matroid then

$$\chi(M) \leq 1 + \max_{N \in R(M)} \left(\min_{C \in C^*(N)} |C| \right) . \tag{7}$$

This is a matroid version of Matula's strengthening of the well known graph result of Szekeres and Wilf. Oxley poses as an unsolved problem whether or not (7) holds for arbitrary loopless or even arbitrary simple binary matroids.

One of the few exact results about chromatic numbers is the following.

THEOREM 2 *A binary matroid has chromatic number* ≤ 2 *if and only if it has no odd circuits.*

This result is the obvious analogue of the easy graph result that $\chi(G) \leq 2$ only when G is bipartite. The matroid version is harder and was first announced by Brylawski and Heron independently in 1972. However now it can be seen that this result is essentially Tutte's 1966 result that the minimal 1-blocks are the odd circuits.

Another general result of Brylawski (1973) is that the class of principal transversal matroids have critical exponent 2 for any field over which they are representable. Principal transversal matroids are an important subclass of transversal matroids, whether there is any similar result for transversal matroids seems open.

Very little is known about the structure of chromatic polynomials. Apart from the fact that their coefficients alternate in sign and are conjectured to be unimodal, the only result of note is that of Stanley (1971) who shows:

If $(a_1, \ldots, a_n)$ is any sequence of positive integers with $a_n = 1$, then there exists a matroid M with

$$P(M;\lambda) = \prod_{i=1}^{n} (\lambda - a_i) \quad . \tag{8}$$

This is just part of a very interesting theory of supersolvable geometric lattices introduced by Stanley. The matroids corresponding to these lattices all have chromatic polynomials with nice factorisation properties.

7. CONCLUSIONS: SOME OPEN PROBLEMS

Apart from the main problem of settling Tutte's tangential block or equivalently 5-flow conjecture there are several other problems here of intrinsic interest.

No progress at all seems to have been made on a characterization of tangential 3-blocks. Based mainly on a lack of counter-examples I conjecture:

There are only 2 tangential 3-blocks, namely $M(K_9)$ and $PG(3,2)$. (1)

Since a proof of this conjecture would include a proof of Hadwiger's conjecture for the case $n = 9$ it is not likely to be easily settled in the affirmative. More promising perhaps is the following problem:

Show that there is no non-graphic tangential 3-block other than $PG(3,2)$. (2)

In other words, show that (1) is equivalent to Hadwiger's conjecture for the case $n = 9$. This is more hopeful for by Jaeger's theorem we know that $P(M;8) > 0$ whenever M is the cocycle matroid of a bridgeless graph and from this it is easy to deduce that there is no cographic tangential 3-block.

An entirely different sort of problem is the following:

When M is representable over GF(q), P(M;q) has, as we have seen, a very nice combinatorial interpretation. Is there any 'reasonable' interpretation of P(M;q) when M is not representable over GF(q) or indeed when M is not representable over any field? (3)

Merton College
University of Oxford

REFERENCES

Appel, K. and Haken, W. (1976). Every planar map is four-colourable. *Bull.Amer.Math.Soc.* 82, 711-12.

Berge, C. (1973). *Graphs and Hypergraphs*. Amsterdam: North Holland.

Birkhoff, G.D. (1912). A determinantal formula for the number of ways of colouring a map. *Annals of Math.* (2), 14, 42-6.

Bixby, R.E. (1977). Kuratowski's and Wagner's theorem for matroids. *J. Combinatorial Theory* (B), 22, 31-53.

Bland, R.G. and Las Vergnas, M. (1978). Orientability of matroids. *J. Combinatorial Theory* (B), 94-123.

Bondy, J.A. and Murty, U.S.R. (1976). *Graph Theory with Applications*. New York: Macmillan.

Brylawski, T.H. (1972). A decomposition for combinatorial geometries. *Trans.Amer.Math.Soc.* 171, 235-82.

Brylawski, T.H. (1975). An affine representation for transversal geometries. *Studies in Appl.Math.* 54, 143-60.

Brylawski, T.H. (1976). A combinatorial perspective on the Radon convexity theorem. *Geometriae Dedicata* 5, 359-66.

Brylawski, T.H. and Lucas, D. (1976). Uniquely representable combinatorial geometries. *Atti dei Convegni Lincei* 17, Tomo 1, 83-104.

Crapo, H.H. (1969). The Tutte polynomial. *Aequationes Math.* 3, 211-29.

Crapo, H.H. and Rota, G.C. (1970). *On the Foundations of Combinatorial Theory: Combinatorial Geometries.* Cambridge Mass.: M.I.T. Press.

Datta, B.T. (1976). On tangential 2-blocks. *Discrete Math.* 15, 1-22.

Datta, B.T. (1976). Non-existence of six-dimensional tangential 2-blocks. *J. Combinatorial Theory* 21, 171-93.

Fiorini, S. and Wilson, R.J. (1978). *Edge-colourings of graphs.* Research Notes in Math. 16. London: Pitman.

Greene, C. (1976). Weight enumeration and the geometry of linear codes. *Studies in Appl.Math.* 55, 119-28.

Heron, A.P. (1972). Matroid polynomials. *Combinatorics: Proc. of the Third British Combinatorial Conf.* (Institute of Math. and its Appl.), ed. D.J.A. Welsh and D.R. Woodall, 164-203.

Jaeger, F. (1976). On nowhere zero flows in multigraphs. *Proc. Fifth British Comb.Conf.*, ed. C.St.J.A. Nash-Williams and J. Sheehan (Utilitas), 373-9.

Jaeger, F. Flows and generalised colouring theorems in graphs. *J. Combinatorial Theory* (B) (to appear).

Las Vergnas, M. Convexity in oriented matroids. *J. Combinatorial Theory* (B) (to appear).

Las Vergnas, M. The Tutte polynomial of a morphism of combinatorial geometries, I. *J. Combinatorial Theory* (A) (to appear).

Las Vergnas, M. (1977). Acyclic and totally cyclic orientations of combinatorial geometries. *Discrete Math.* 20, 51-61.

Lindström, B. (1978). On the chromatic number of regular matroids. *J. Combinatorial Theory* (B), 24, 367-9.

Martin, P. (1978). Remarkable valuation of the dichromatic polynomial of planar multigraphs. *J. Combinatorial*

Theory (B), 24, 318-24.

Matula, D. (1972). k-components, clusters and slicings in graphs. *SIAM J.Appl.Math.* 22, 459-80.

Minty, G.J. (1967). A theorem on three colouring the edges of a trivalent graph. *J. Combinatorial Theory* 2, 164-7.

Oxley, J.G. (1978). Colouring, packing and the critical problem. *Quart.J.Math. Oxford* (2), 29, 11-22.

Oxley, J.G. On a covering problem of Mullin and Stanton for binary matroids. *Aequationes Math.* (to appear).

Oxley, J.G. and Welsh, D.J.A. The Tutte polynomial and percolation. *Graph Theory and Related Topics.* Academic Press (to appear).

Rosenstiehl, P. and Read, R.C. (1978). On the principal edge tripartition of a graph. *Ann. Discrete Math.* 3, 195-26.

Rota, G.C. (1964). On the foundations of combinatorial theory, I. *Z.Wahrsch.* 2, 340-68.

Stanley, R. (1971). Modular elements of geometric lattices. *Alg.Univ.* 1, 214-17.

Stanley, R. (1972). Supersolvable lattices. *Alg.Univ.* 2, 197-217.

Stanley, R. (1973). Acyclic orientations of graphs. *Discrete Math.* 5,

Seymour, P.D. Decomposition of regular matroids. *J. Combinatorial Theory* (to appear).

Seymour, P.D. On multicolourings of cubic graphs and conjectures on Fulkerson and Tutte. *Proc. London Math.Soc.* (to appear).

Szekeres, G. and Wilf, M.S. (1968). An inequality for the chromatic number of a graph. *J. Combinatorial Theory* 4, 1-3.

Tutte, W.T. (1954). A contribution to the theory of chromatic polynomials. *Canad.J.Math.* 6, 80-91.

Tutte, W.T. (1966). On the algebraic theory of graph

colourings. *J. Combinatorial Theory* 1, 15-50.

Tutte, W.T. (1969). A geometrical version of the four colour problem. *Combinatorial Math. and its Applications*, ed. R.C. Bose and T.A. Dowling, Univ. of North Carolina Press, Chapel Hill, 553-61.

Veblen, O. (1912). An application of modular equations in analysis situs. *Ann.Math.* 14, 86-94.

Wagner, K. (1964). Beweis einer Abschwächung der Hadwiger-Vermutung. *Math.Ann.* 153, 139-41.

Welsh, D.J.A. (1976). *Matroid Theory.* London Math.Soc. Monograph No.8. London: Academic Press.

Zaslavsky, T.K. (1975). Facing up to arrangements for face-count formulas for partitions of space by hyperplanes. *Mem.Amer.Math.Soc.* 154.

Index